MIGHTY CHAMPIONS OF MENTAL HEALTH EDUCATION

Dr. Shivam Dubey & Dr. Salil K. Gupta

About the Author

Shivam Dubey, MD, FAPA is a Harvard-trained psychiatrist, global mental health advocate, and founder of **Mental Health Education Inc.**, a nonprofit transforming how mental health is taught to children, families, and educators around the world.

With a deep passion for neuroscience, storytelling, and education, Dr. Dubey has authored and co-authored numerous books blending evidence-based science with emotional literacy—including the Turbo Sparky series and The Adventures of Aqualand, which teach mental health and social-emotional learning through engaging, age-appropriate stories.

As the architect of the **Mighty Champions of Mental Health Program**, Dr. Dubey is building a global movement to empower 1 million individuals with practical mental health skills rooted in neuroplasticity, mindfulness, and compassion. His curriculum is currently used in over 30 schools and has reached underserved communities across the U.S., India, and beyond—including free sessions supported by the NBA's Orlando Magic.

Whether writing, teaching, or working with families, Dr. Dubey's mission remains the same: to make mental health education accessible, scientific, and deeply human—one child, one parent, one story at a time.

Salil K. Gupta, MD, neonatologist (retired) at Advent Health, has been recognized by Marquis Who's Who Top Doctors for dedication, achievements, and leadership in health care.

With nearly four and a half decades of experience to his credit, Dr. Gupta has established a successful career in health care. Before his retirement in 2020, he served as a neonatologist at Advent Health since 2018. Prior to this, he was a neonatologist and medical director at Envision Healthcare from 2013 to 2018. His earlier roles include serving as a newborn specialist, neonatologist, and medical director of newborn services at St. Elizabeth's Hospital with the HSHS Hospital Sisters Health System from 1993 to 2013. He was also chief medical officer at St. Elizabeth's Hospital from 2008 to 2010 and a neonatologist and medical director of newborn services at Scott Air Force Base from 1990 to 1993. Laying a solid educational foundation, he earned an MBBS from Maulana Azad Medical School in India in 1979, an undergraduate degree awarded to students who have completed a medical education program, and a Doctor of Medicine in obstetrics and gynecology from the same institution in 1982. He went on to complete his residency in pediatrics at The University of Iowa in 1987 and a fellowship in perinatal neonatal medicine at The University of Iowa in 1990.

Post-retirement, Dr. Gupta has dedicated his time as locum tenens with Envision Healthcare. Although he is no longer in clinical practice, he remains actively involved through volunteer work and various service projects within his field. Previously, he taught family practice residents at the University of Central Florida and continues to mentor and counsel young doctors. Additionally, he has dedicated his time as a medical service officer with Matthew's Hope Ministries since 2021. He is also a volunteer teaching faculty member at the University of Central Florida, is involved in service projects with the Rotary Club of

Winter Garden, and serves as a board member for the March of Dimes and Matthew's Hope. In light of his impressive undertakings, he was honored with a Distinguished Service Award from the Scrivner Foundation in 2010 and Excellence in Pediatric Care from the Illinois Department of Public Health in 2006.

In the coming years, Dr. Gupta's main focus is on contributing to the welfare of humanity and giving back to the community and the world that has given so much to him. He expresses gratitude to God for this abundance and is actively involved in community service. He is dedicated to Matthews Hope, a homeless service center, where he serves as a full-time medical service officer on a voluntary basis. He has successfully established the medical home and has recruited nurse practitioners, nurses, and numerous volunteers to support the initiative. After practicing as a resident physician for 10 years and as a neonatologist for 30 years until his retirement in September 2020, he is now fully committed to his volunteer work.

Acknowledgment

Thanks to Krishiv Mittal and Raghav
Bhatnagar for their powerful contributions to the
mission of Mental Health Education for Teenagers.

Join them on TheRaghavPodcast—where Raghav
and Krishiv host conversations with mental health
experts. Available now on Spotify and Apple
Podcast.

Spotify

Apple Podcast

Before You Begin

Dear parents, teachers, and students,

At Mental Health Education, we are invested in the mental well-being and future success of every child and teenager. Here are the benefits of this book and how the Mighty Champions program works.

BENEFITS AND HOW THIS PROGRAM WORKS

Before you begin, here's what this program can do for your future and how the Mighty Champions process works.

1. Career and college benefits
- Shows leadership on applications and resumes
- Provides verified community-service hours
- Demonstrates emotional intelligence, which colleges and employers value
- Offers a certification that stands out among extracurricular activities
- Builds confidence, communication skills, and real problem-solving ability
- Helps students manage stress in school, sports, and future jobs
- Increases eligibility for leadership roles, clubs, and scholarships
- Allows participation in research activities such as posters, short papers, and presentations at scientific conferences around the world

- All activities require registration through our program so we can verify and track the hours spent

2. What this program teaches

This book trains students in ten mental-health competencies that support success in life:

self-awareness, self-management, social awareness, relationship skills, responsible decision-making, problem solving, CBT tools, positive psychology, mindfulness, and resilience.

3. How the learning works

- Each chapter teaches one competency using science and real examples
- Students complete activities that build real-life experience
- Quizzes help check understanding
- Reflection questions connect the skills to daily life
- Students practice the skills by teaching or sharing them with others

4. How you earn community-service hours

Students can complete volunteer hours by:

- Reading a chapter to a younger child or a group
- Teaching one skill to classmates or siblings
- Leading a mindfulness minute or discussion
- Starting or joining a Mental Health Champion Club
- Helping with school awareness projects
- Participating in research, podcast activities, or youth events

All volunteer work must be registered in advance so hours can be officially verified.

5. How to become a certified Mighty Champion

- **Step 1:** Read each chapter
- **Step 2:** Complete the reflection activities
- **Step 3:** Finish all the quizzes
- **Step 4:** Complete at least 1 registered volunteer-service hour
- **Step 5:** Register and take the final exam
- **Step 6:** Receive your Mighty Champion Certificate and transcript

6. What you can do after certification

- Add your certification and transcript to your resume or CV
- Use verified hours for school requirements
- Lead projects and support younger students
- Participate in research presentations, events, and the podcast
- Join a growing community of students promoting mental health

This book is your step-by-step guide to becoming emotionally stronger, more confident, and well prepared for your academic, personal, and professional future.

Contents

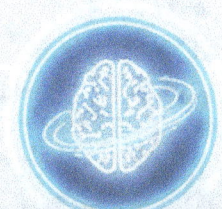

CHAPTER 1

SELF-AWARENESS

Know Yourself, Grow Yourself

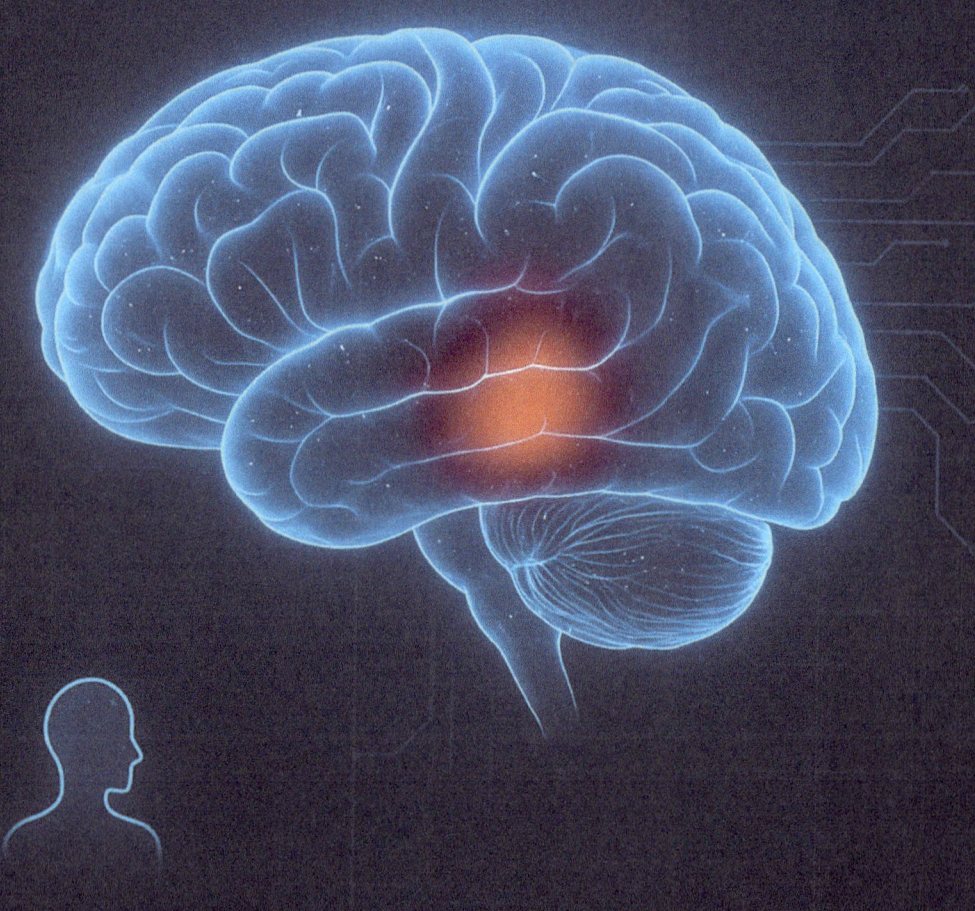

Section 1:

Biology– Your Brain, Your Blueprint

Have you ever said something in the heat of the moment and then thought, "Why did I do that?" That reaction isn't just about emotions—it's about how your brain is built to protect you.

Let's break it down.

Meet Your Brain's Three-Part Team

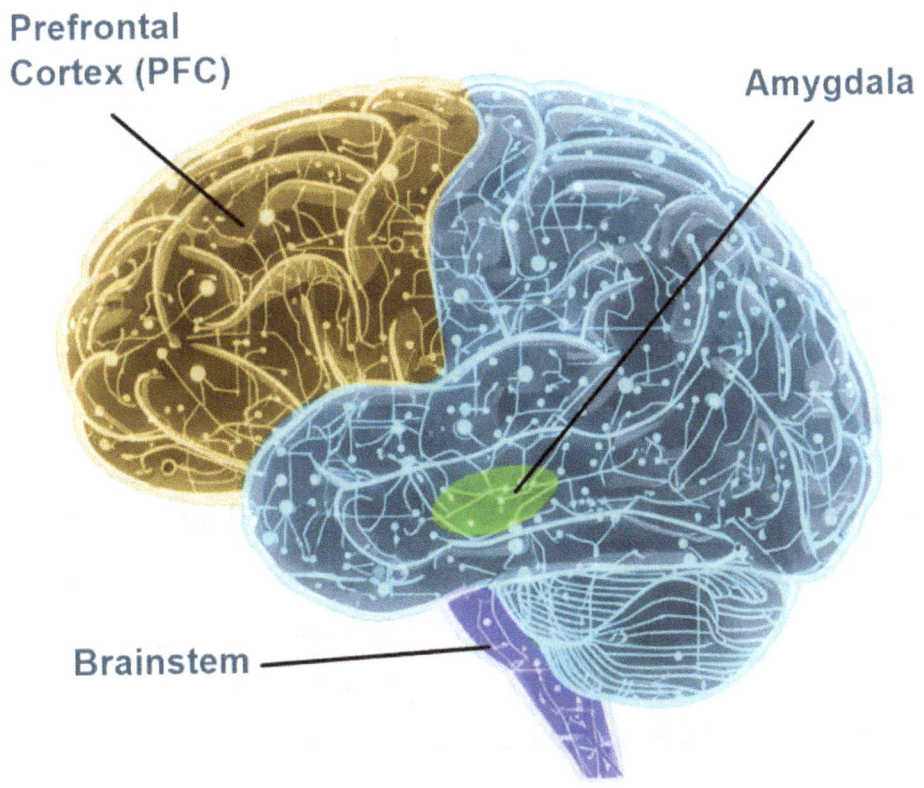

Prefrontal Cortex (PFC)

Amygdala

Brainstem

Brainstem – Your Emergency Responder

This is the oldest part of your brain. It handles automatic functions like breathing and heart rate. When you feel scared, left out, or in danger, this part kicks in without asking you. It makes you fight, run, or freeze to survive. It's like the emergency button.

Amygdala – Your Emotional Alarm System

This small, almond-shaped part constantly scans the world for anything that might hurt or embarrass you. If someone rolls their eyes or yells at you, your amygdala sets off the alarm. It sends stress chemicals like adrenaline and cortisol—making your heart pound, your face flush, or your voice shake. It reacts before your thinking brain can even speak up. That's why you may shout, shut down, or storm off before you even realize it.

Prefrontal Cortex (PFC) – Your Inner Coach

Located right behind your forehead, this part helps you pause, plan, and make smart choices. It helps you breathe through big feelings, reflect on what matters, and stay in control. But here's the catch—it's the last part of your brain to fully develop (usually in your mid-20s). That means teenagers often feel intense emotions but haven't fully built the brakes yet.

Why This Matters

If no one explains how your brain works, it can feel like something's wrong with you. But nothing is wrong. You're not broken. You just haven't been taught to train your brain. When you practice things like breathing, pausing, or journaling, you build the prefrontal cortex—and weaken the grip of the amygdala.
This is how you become the boss of your brain.

DNA, Genes & Emotional Sensitivity

Let's talk about where you come from. Your DNA is like the instruction manual passed down from your parents and ancestors. It tells your body things like: "Give her curly hair," "Make his nose like Grandpa's," "Pass on Mom's green eyes."

But DNA also carries more invisible things—like how easily you might get anxious or how sensitive you are to stress. If your parents or grandparents had depression, ADHD, or mood issues, that tendency might live in you too. This is called genetic predisposition.

But guess what? Genes are like switches. You can't change the manual—but your choices and environment can turn certain switches on or off. This is called epigenetics—the science of how your lifestyle (stress, sleep, habits, food, love, thoughts) shapes how your genes behave.

So yes, you might carry emotional wiring—but with awareness, you can shape how it plays out.

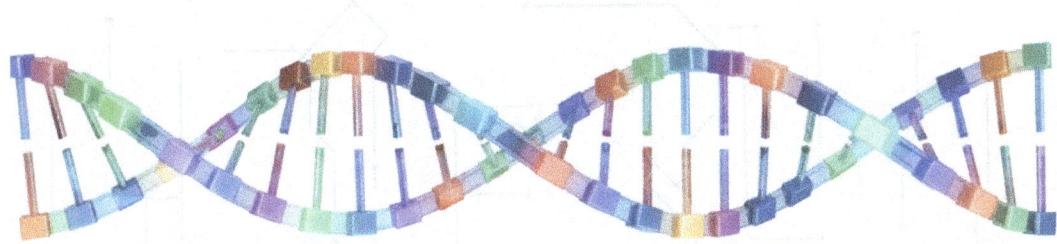

Example:

* You might have a gene for being extra sensitive.

* But if you learn how to recognize stress early, practice calming tools, and build positive relationships, that sensitivity can become a gift—like noticing others' emotions and being a great friend or artist.

Trivia

Your Mood Has a Genetic Signature

• **SLC6A4** – Also known as the serotonin transporter gene. A shortened version of this gene is linked to higher sensitivity to stress and greater risk of depression.

• **BDNF** – Brain-Derived Neurotrophic Factor. This gene affects how your brain grows and adapts. It's essential for learning, memory, and recovery from trauma.

• **DRD4** – A dopamine receptor gene linked to novelty-seeking behavior and ADHD symptoms.

• **COMT** – Influences how quickly you clear stress chemicals like dopamine. Some versions make you more chill, others more reactive.

These aren't "bad" genes—they're **blueprints with potential.** With the right habits, love, and
emotional skills, you can shape how these genes express themselves.

Section 2:

Psychology– The Invisible Scripts of Your Life

By the time you're a teenager, you've developed a mental operating system—a set of automatic beliefs and reactions that run your daily life. Some are helpful. Some hold you back.

We call these core beliefs.

What Are Core Beliefs?

Core beliefs are like invisible "rules" your brain picked up without you knowing. They help you survive, make sense of things, and decide who you are.

They come from:

• Repeated messages from parents, teachers, or social media.

• Things people said to you when you were little (good or bad).

• Experiences—like being left out, praised, teased, or ignored.

Your brain turns these moments into deep beliefs:

• "I always mess up."

* "I'm only lovable if I'm perfect."

* "I must please people to stay safe."

These beliefs run like background apps on your phone—they drain energy even when you're not using them.

They shape how you:

* Talk to yourself

* React in relationships

* Handle failure

* View your potential

How Core Beliefs Are Like Riding a Bike?

When you learned to ride a bike, your body developed automatic balance reactions. You don't think, "Lean left, now right!"—your brain just does it. Core beliefs are the same: they become automatic.

But what if you learned to ride a bike wrong—and now you tip over often? You'd have to retrain your muscles. Same goes for your mind.

Rewriting Core Beliefs: Step-by-Step

1. Awareness – "Wait... what did I just tell myself?"

2. Inquiry – "Is that true? Where did I first hear that?"

3. Replace – "What would be a more helpful belief?"

4. Repeat – Keep practicing the new one. Your brain will rewire.

This is called neuroplasticity—your brain's power to reprogram itself.

Examples:

- **Old:** "If I fail, I'm worthless." – **New:** "If I fail, I learn something."

- **Old:** "People always leave." – **New:** "Some people do, but others stay. I'm worthy of love."

The more you practice this, the more your self-awareness becomes a superpower.

Section 3:

Social Influences – The Mirror World Around You

Humans are wired to connect. From the moment you're born, your brain watches others to learn how to act, feel, and think. This is possible because of mirror neurons—special brain cells that copy what we see.

That's why you yawn when someone yawns—or feel nervous when your friend is anxious.

But it goes deeper:

• You start to copy emotional habits of your family or culture.

• You absorb what your community says is normal, beautiful, successful, or lovable.

Social Influences That Shape You:

• Family rules: "Boys don't cry." "Don't talk back." "Be strong."

• Peer pressure: "Everyone's doing it." "Don't be weird."

• Social media: "They're so perfect." "I'm not enough."

• Cultural beliefs: "You have to succeed or you're a failure."

Without self-awareness, these influences program you—just like an app installs on your phone.

But with self-awareness, you get to ask:

- "Do I even agree with this?"

- "Does this belief help me—or hold me back?"

Ask Yourself:

- "Am I doing this because I want to—or to fit in?"

- "What would I do if no one was watching?"

- "Does this version of me feel true—or just safe?"

Self-awareness gives you the power to choose:

- What thoughts to believe?

- What patterns to break?

- What kind of life to build?

And that's the beginning of true freedom.

Section 4:

Metacognition – The Mystery Within

There is a fascinating ability your brain possesses: the power to think about your own thoughts. This is called metacognition.

Metacognition allows you to monitor, analyze, and regulate your internal experience. Instead of reacting automatically to emotions or thoughts, you can pause, reflect, and choose your response—as if stepping outside your experience to examine it from a higher perspective.

The Science Behind It:

Although there isn't one single "metacognition center," scientists believe this capacity arises from a network of brain regions—especially the Default Mode Network (DMN). The DMN becomes active when you're not focused on the outside world—during daydreaming, recalling memories, or self-reflection.

It includes areas like:

• The medial prefrontal cortex, which is involved in thinking about the self.

• The posterior cingulate cortex, which evaluates relevance and focus.

• The angular gyrus and other association areas.

This network helps you ask questions like:

• What am I feeling right now?

• Why did I act that way?

• Is that thought really true?

Metacognition is like having an internal feedback system—one that can monitor your mental state in real time and help guide your decisions with insight instead of impulse.

What Metacognition Looks Like in Action?

• **Recognizing**: I feel angry right now. My heart is racing. My fists are clenched.

• **Pausing**: I want to yell... but is that the best option?

• **Analyzing**: Where did this belief come from that I always have to defend myself?

• **Choosing**: I'm going to take a breath before responding.

These are not abstract ideas—they are trainable skills. Practicing metacognition builds emotional intelligence, resilience, and mental flexibility.

Try This Brain Training Experiment

The next time you feel a strong emotion, pause for a moment and narrate the situation like a scientist observing a subject:

The subject is experiencing an increased heart rate and shallow breathing. The trigger was identified as social embarrassment. The brain is preparing a reactive response. Initiating conscious delay to allow prefrontal cortex input.

It may sound humorous—but it actually activates your reflective circuits and trains your brain to respond thoughtfully.

Why This Matters?

When you develop metacognition:

• You reduce impulsive reactions.

• You gain emotional clarity.

• You separate your identity from passing thoughts and feelings.

The goal is not to eliminate emotions, but to create mental space around them—so you are no longer controlled by them.

This is the foundation of advanced self-awareness and self-mastery.

Coming Next: Self-Management

Now that you've mapped your inner world, it's time to take charge.

In Chapter 2, we'll explore:

• How to calm emotional storms

• Simple daily habits to build emotional strength.

• Why dopamine makes things harder—and how to use it wisely

You're not just becoming self-aware. You're becoming a mighty champion of your mind.

Chapter 1 Quiz:

How Self-Aware Are You?

Instructions:

Answer each question honestly. Choose the best option that reflects your experience. At the end, check your answers and explanations to understand your current level of self-awareness.

Part 1: Biology and Brain Function

1. What is the main function of the amygdala?
A) Planning your day
B) Managing your heartbeat
C) Scanning for threats and triggering emotions
D) Helping with math skills

2. Which part of the brain helps you make thoughtful decisions and manage impulses?
A) Brainstem
B) Amygdala
C) Prefrontal Cortex
D) Hippocampus

3. What does the brainstem control?
A) Social behavior
B) Breathing and heart rate
C) Imagination
D) Speech

4. What is neuroplasticity?

A) The brain's refusal to change
B) The way your brain gets heavier with age
C) The brain's ability to change based on experience and repetition
D) A part of the skull

5. Which statement about genes is true?

A) They cannot be changed in any way
B) You're stuck with whatever emotions your genes give you
C) Your behavior and environment can influence how genes are expressed
D) Only medication can affect gene expression

Part 2: Psychology– Thoughts and Beliefs

6. What are core beliefs?

A) Random thoughts you have once in a while
B) Deep unconscious rules you live by
C) Scientific theories you believe in
D) Opinions you read on social media

7. Which of the following is an example of a harmful core belief?

A) "Mistakes are part of learning."
B) "I need to be perfect to be loved."
C) "I'm growing every day."
D) "I can handle tough emotions."

8. How can you change a harmful core belief?
A) Ignore it and hope it goes away
B) Repeat it until you believe it more
C) Notice it, question it, and replace it with a healthier one
D) Ask others to tell you what to believe

Part 3: Social Awareness

9. What are mirror neurons responsible for?
A) Reflecting light
B) Making you copy and empathize with others
C) Creating dreams
D) Helping your eyes move

10. Which of the following shows social self-awareness?
A) Following a trend without thinking
B) Asking, "Does this reflect who I really am?"
C) Trying to fit in at all costs
D) Doing what your friends do even when it feels wrong

Part 4: Metacognition

11. What is the Metacognition part of you?
A) Your teacher at school
B) A fantasy character
C) The part of your mind that can watch your emotions and thoughts without getting caught in them
D) Your amygdala in disguise

12. Why is Metacognition important for self-awareness?

A) It helps you avoid all emotions

B) It tells you what's wrong with other people

C) It gives you space to choose your response

D) It gives you more emotions

Answers

1-C)

2-C)

3-B)

4-C)

5-C)

6-B)

7-B)

8-C)

9-B)

10-B)

11-C)

12-C)

Scoring Guide

Count how many you got right:

• **10–12 correct: Self-Awareness Superhero**
You understand your brain, beliefs, and emotions deeply. Keep practicing your skills.

• **7–9 correct: Awareness Explorer**
You're getting the hang of it. Stay curious and keep reflecting.

• **4–6 correct: Just Getting Started**
Great beginning. Reread some sections and try the self-experiments.

• **0–3 correct: Welcome to the Journey**
Everyone starts somewhere. You're not behind—you're just at the start of your growth.

Self-Reflection:

Questions to Know Yourself, Grow Yourself

1. What emotion do I feel the most lately—and when does it usually show up?

2. What do I do when I feel angry, anxious, or hurt? (Do I talk, shut down, scroll, eat, avoid, or explode?)

3. What's one thing I often tell myself that might not be true?

4. When do I feel most like myself—relaxed, confident, or joyful?

5. What's one habit I've picked up from others that I'm not sure I want to keep?

6. Who in my life helps me feel safe, supported, and understood?

7. What's one small change I can make to feel better emotionally every day?

8. If I could pause time during a stressful moment, what would I do differently?

NOTES

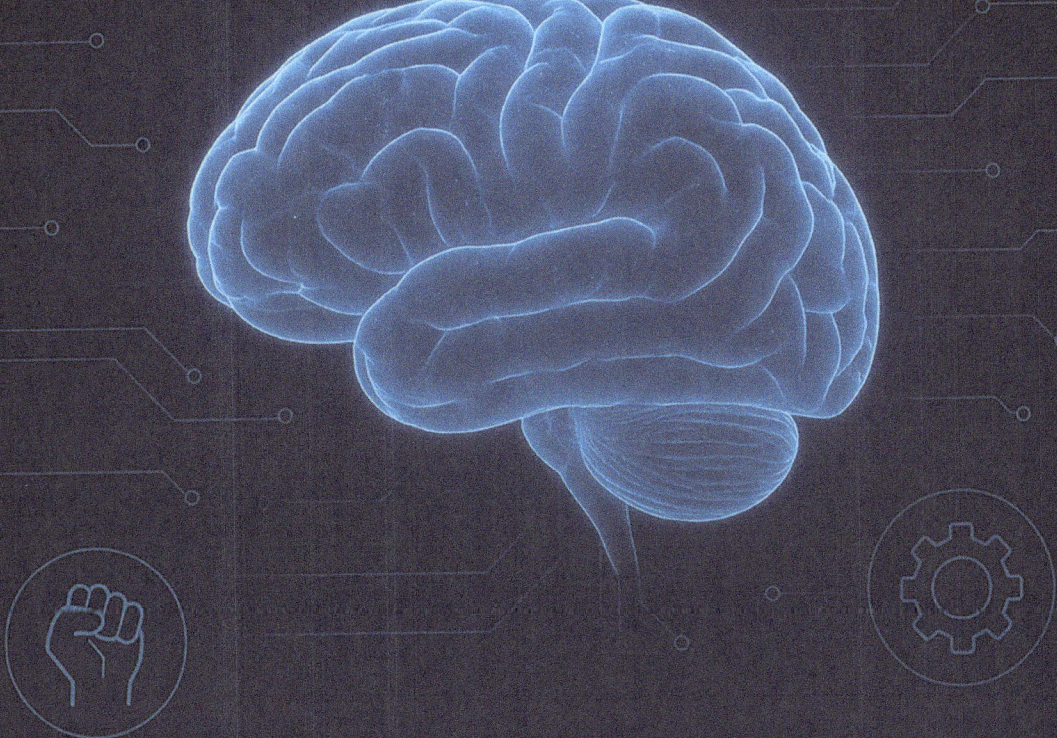

CLASSIFIED RESEARCH

SELF-MANAGEMENT

Be the Boss of Your Reactions

Section 1:

What Is Self-Management

Self-awareness is noticing what you feel.

Self-management is choosing what you do next.

It's the skill of handling your emotions, thoughts, and behaviors—especially when things feel out of control.

When you're sad, angry, scared, or even excited—your body and brain get flooded with chemicals.

Your job is to press the pause button long enough to respond instead of react.

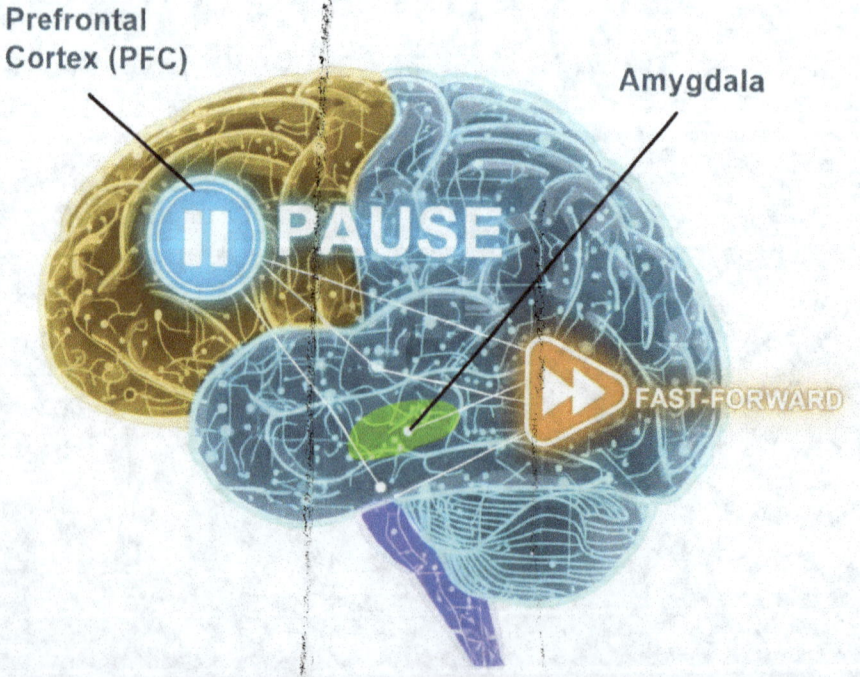

Brain-Body Trivia

The Chemistry of Emotions

When you feel strong emotions like fear, sadness, anger, or excitement, your brain activates the limbic system—especially the amygdala. This triggers a fast release of neurochemicals that prepare your body to respond.

• Adrenaline (epinephrine) increases heart rate and energy for quick action.

• Cortisol helps manage stress but can harm your body if it stays high for too long.

• Dopamine drives motivation and reward-seeking behavior.

• Serotonin helps stabilize mood and supports emotional balance.

These chemicals create the physical sensations of emotions—like a racing heart, tense muscles, or butterflies in your stomach—proving that emotions are both mental and biological experiences.

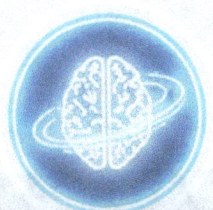

Section 2:

The Power of the Pause

You can't stop emotions from coming. But you can stop yourself from doing something you regret.

That's where the pause comes in. When you pause:

• Your prefrontal cortex (PFC) turns on

• Your amygdala quiets down

• You get time to think

Try the "5-Second Pause"

When you feel triggered:

1. Notice the emotion
2. Take a slow breath in (5 seconds)
3. Say: "I'm feeling ___"
4. Ask: "What do I want to do next?"
5. Then choose your response

This rewires your brain over time.

What Does "Rewire Your Brain" Mean?

Your brain has something called neuroplasticity. That means it can physically change and build new connections based on what you practice. Think of your brain like a forest full of paths. Every time you repeat an action—like breathing calmly instead of yelling—you walk a new path. The more often you do it, the clearer and faster that path becomes.

This is exactly how you learned to do other things.

When you first learned to ride a bike or swim, it felt strange and hard. You had to think about every little move. But the more you practiced, the more automatic it became. That's because your brain was wiring itself to do those actions smoothly.

The same thing happens with emotional skills.

If you practice the 5-second pause over and over, your brain starts building a calm, thoughtful path instead of always choosing panic or anger. Over time, your new reaction becomes natural.

So, rewiring your brain simply means: practicing something new until your brain adopts it as your default. It's like training a new reflex—one that helps you stay in control.

Section 3:

Tools to Manage Emotions

Self-management isn't one big skill—it's a toolbox.

Here are five tools anyone can use:

1. Name It to Tame It:
Say your emotion out loud or write it down. Just naming your feeling reduces its power.

2. Move Your Body:
Walk, stretch, dance, or shake it out. Physical movement resets your nervous system.

3. Breathe Slowly:

Try this:
- Breathe in for 4 seconds
- Hold for 4 seconds
- Breathe out for 6 seconds

Repeat 3 times. Your brain will thank you.

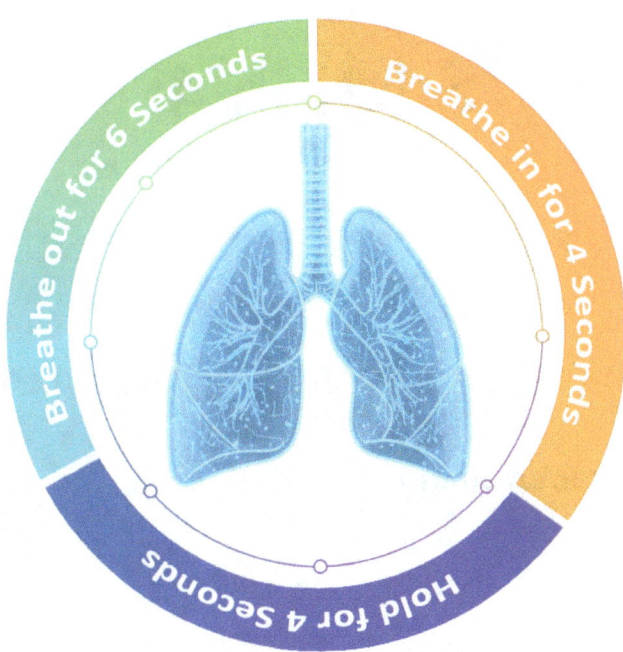

Follow-Along Breathing Exercise

4. Change the Scene:
Step outside. Splash cold water. Go somewhere new for a few minutes. Change your environment to change your state.

5. Talk to a Trusted Person
Talking helps your brain organize the emotion. Choose someone who listens well.

Section 4:

What's Happening in Your Brain

When you use these tools, you're not just calming down—

You're doing brain training.

Each time you pause instead of exploding...

Each time you name your emotion instead of suppress it...

Each time you breathe instead of breaking something...

...you strengthen the neural connection between your prefrontal cortex and amygdala.

That's how your emotional-regulation circuits grow stronger. It's like lifting weights for your mind.

Section 5:

Dopamine vs. Discipline

Your brain loves quick rewards: sugar, scrolling, drama, attention.

That's dopamine talking.

But real peace comes from building discipline—choosing what's right over what's easy.

Every time you:

- Stay calm in conflict

- Keep going when it's hard

- Get off your phone and do what matters...

...you're building your Champion brain.

Dopamine fades fast. Discipline builds strength.

Section 6:

Visualize the Future You

Close your eyes. Imagine a version of you who:

• **Handles stress calmly**

• **Knows how to pause and breathe**

• **Makes good choices, even in hard moments**

That person is already inside you—waiting to be trained.
Every time you manage your emotions, you get one step closer.

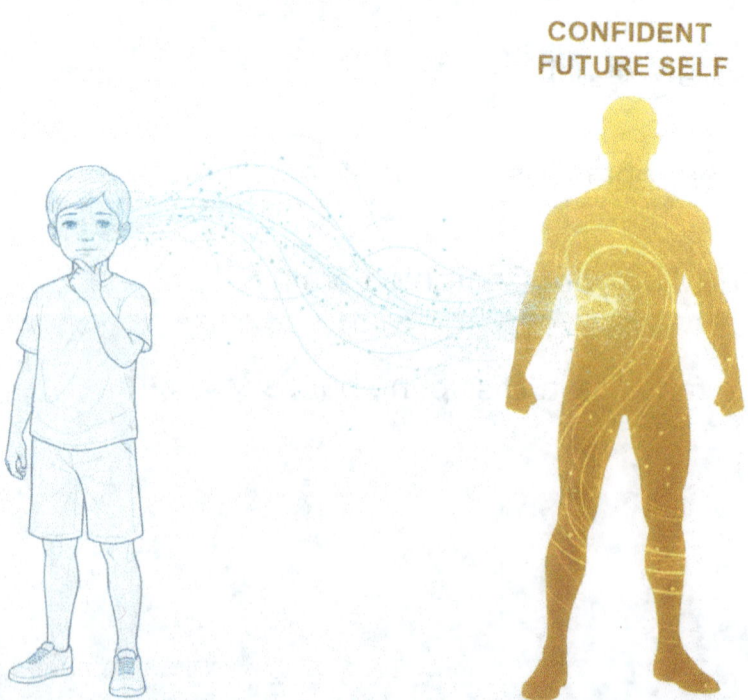

CONFIDENT
FUTURE SELF

Coming Next: Social Awareness

In Chapter 3, you'll learn how to understand others—not just yourself.

We'll explore:

* Empathy

* Respecting differences

* How to build strong relationships

Because being mentally strong means helping others feel safe, too.

Chapter 2 Quiz:

How Well Do You Manage Your Reactions?

Instructions:
Choose the best answer (A, B, or C).

1. What is self-management?
A) Hiding your feelings
B) Choosing how to respond after you feel something
C) Controlling others' emotions

2. What happens in your brain when you pause before reacting?
A) The amygdala gets louder
B) The prefrontal cortex takes over and helps you think
C) Your brain shuts down

3. Which of these is NOT a self-management tool?
A) Blaming someone
B) Breathing slowly
C) Naming the emotion

4. What does "Name it to tame it" mean?
A) Say how someone else made you feel
B) Give your emotion a funny nickname
C) Label your emotion to calm your brain down

5. What's the difference between dopamine and discipline?
A) Dopamine is fast pleasure; discipline builds strength over time
B) Dopamine is always bad
C) Discipline is just for athletes

Answers

1-Correct answer: B
Explanation: Self-management means choosing what you do after an emotion shows up.

2-Correct answer: B
Explanation: Pausing helps the thinking part of the brain (PFC) quiet down the emotional center (amygdala).

3-Correct answer: A
Explanation: Blaming others doesn't help you manage your-self—it's a reaction, not a solution.

4-Correct answer: C
Explanation: Naming an emotion reduces its intensity in your brain. It helps the PFC take
control.

5-Correct answer: A
Explanation: Dopamine gives quick reward, but discipline gives long-term peace and power.

Scoring Guide

Count how many you got right:

5 correct: Emotion Ninja
You know how to manage your reactions like a pro.

3–4 correct: Emotion Explorer
You're learning and building strength.

0–2 correct: Just Getting Started
Every skill starts with learning. Keep going—you're on the right path.

Self-Reflection Questions:

Use these to think deeper and apply what you've learned.

1. What is one emotion I feel often that's hard to manage?

2. What do I usually do when I feel that emotion—and does it help or hurt?

3. When was the last time I paused before reacting? What happened?

4. What tool from the chapter would help me most right now? (Breathing, moving, naming it, talking, changing the scene)

5. What's a situation where I often react quickly—and what would "the future me" do differently?

6. What is one habit I want to build to stay calm when things get stressful?

7. Who can I talk to when I need help managing tough feelings?

8. If I practiced self-management every day for 30 days, how would my life feel different?

NOTES

CHAPTER 3
SOCIAL AWARENESS

Understand Others, Build Stronger Connections

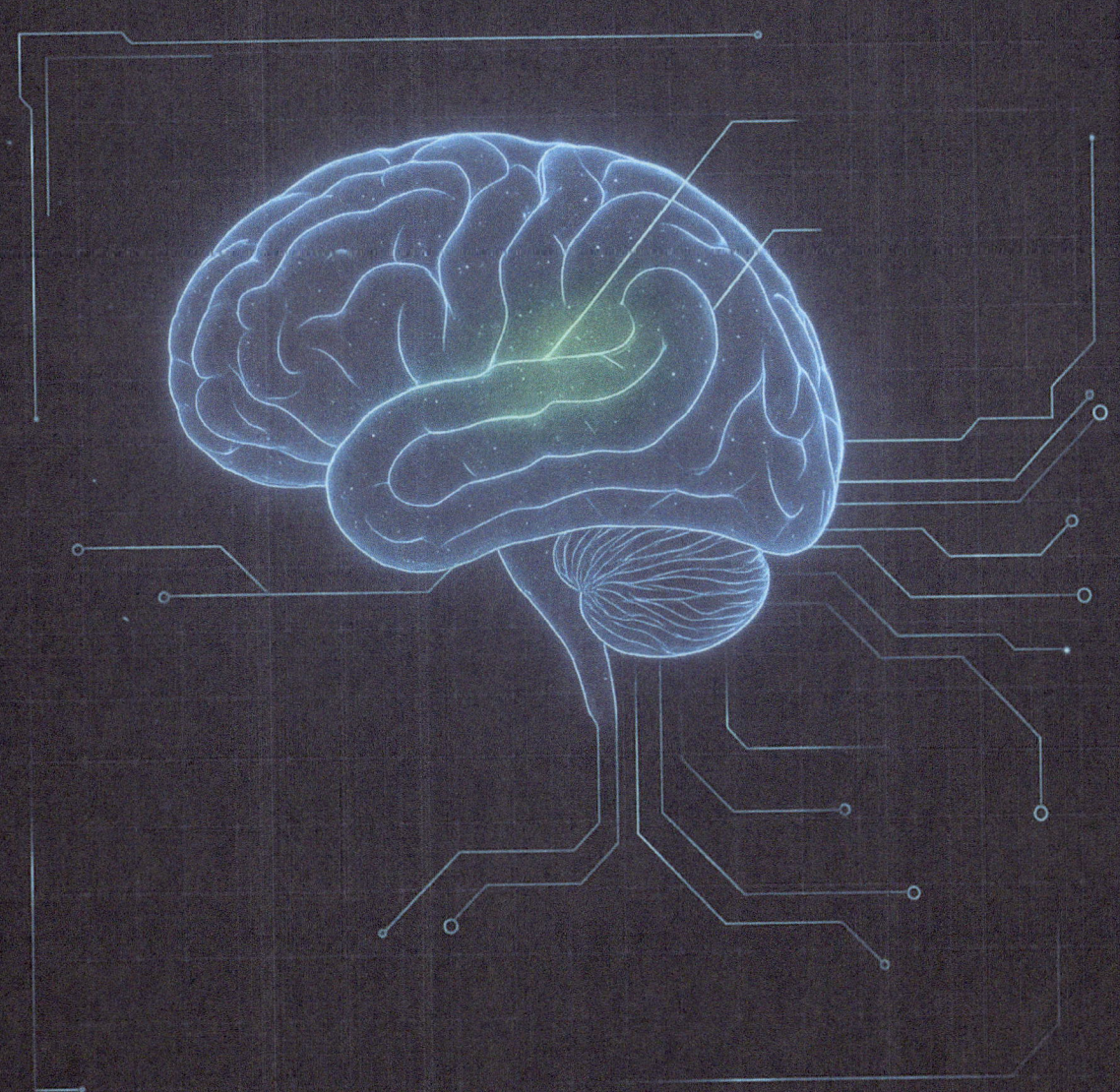

Section 1:

What Is Social Awareness

Social awareness means being able to:

- Understand other people's feelings

- Notice what's happening around you

- Respect differences in background, belief, and experience

It's the skill that helps you build better relationships—with friends, family, classmates, teachers, and people you don't even know.

It begins with one powerful word: empathy.

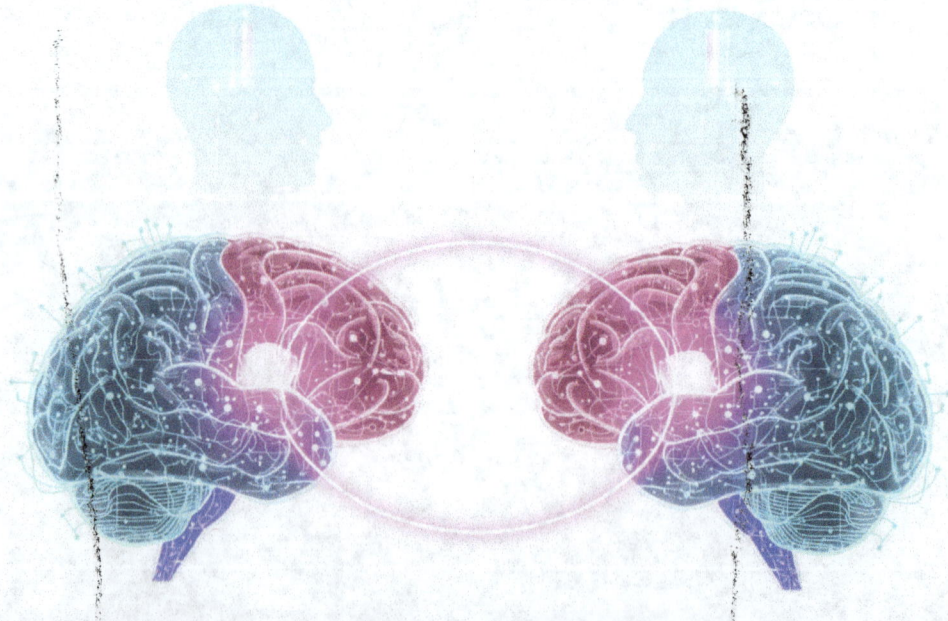

Section 2:

Empathy – Feel What Others Feel

Empathy is when you can imagine what it's like to be someone else—even if you haven't been through the same thing.

There are two types:

• **Emotional Empathy**: You feel what the other person feels. If your friend is crying, you feel sad too.

• **Cognitive Empathy**: You understand what the other person might be feeling, even if you don't feel it yourself.

Why Empathy Matters:

• It helps you connect instead of judge

• It makes people feel seen, safe, and supported

• It reduces conflict and builds trust

You don't need to have all the answers. You just need to be present.

Section 3:

Respecting Differences

Every person you meet has a different story.

Different family, different beliefs, different culture, different brain.

You don't have to agree with everyone.

But you do need to respect their human dignity.

What This Looks Like:

• Listening without interrupting.

• Noticing when someone is being left out.

• Learning about cultures, religions, or abilities that are different from yours.

• Avoiding jokes or labels that hurt others
Understanding isn't about changing your values. It's about expanding your heart.

Section 4:

Reading Social Cues

Social awareness also means reading non-verbal signs:

- **Body language**

- **Facial expressions**

- **Voice tone**

People don't always say what they feel. But their body often shows it.

Try This:
When talking to someone, ask:

- Do they seem open or closed off?

- Are they making eye contact or avoiding it?

- Is their voice calm, shaky, fast, or flat?

These clues can help you respond with kindness and care.

Section 5:

The Power of Small Actions

You don't need a big speech to make someone feel seen.

Small actions matter:

• A smile

• A genuine "How are you?"

• Including someone who looks alone

• Saying "thank you" or "I understand"

These moments create what scientists call micro-connections—tiny moments that build trust and belonging.

When you show others they matter, your own brain lights up too.

Section 6:

How It Changes Your Life

When you practice social awareness:

• Your friendships get stronger

• You become a better team player

• People trust you more

• You feel less lonely and more connected

Empathy isn't weakness—it's your superpower for emotional intelligence and leadership.

Trivia

Mirror Neurons Make You Empathic

• Mirror neurons in your brain activate when you see someone else smile, cry, or yawn—helping you "mirror" their emotion and feel what they feel.

• The insula and anterior cingulate cortex (ACC) light up when you experience empathy—same areas that process pain. That's why seeing someone hurt can make you wince!

• Emotional intelligence studies show that teens with higher social awareness are more likely to succeed in leadership, team sports, and friendships.

Coming Next: Relationship Skills

In Chapter 4, we'll explore:

• How to make and keep good friends

• How to resolve conflicts peacefully

• How to communicate clearly and kindly

Because understanding others is just the beginning—building strong, healthy relationships takes skill too.

Chapter 3 Quiz:

How Socially Aware Are You?

Instructions:
Choose the best answer (A, B, or C).

1. What is social awareness?
A) Controlling how others feel
B) Understanding your own emotions only
C) Noticing how others feel and what's happening around you

2. What does empathy mean?
A) Feeling sorry for someone
B) Imagining what someone else is feeling or going through
C) Giving advice to fix someone's problem

3. Which of these shows emotional empathy?
A) Understanding why your friend is sad, even if you don't feel sad
B) Feeling sad yourself because your friend is crying
C) Telling your friend to cheer up immediately

4. What's one way to show respect for someone different from you?
A) Correct them until they agree with you
B) Stay silent and ignore them
C) Listen to their story without interrupting or judging

5. What is a "micro-connection"?

A) A small moment of genuine human connection
B) A mistake in technology
C) A tiny argument with a friend

Answers

1-Correct answer: C
Explanation: Social awareness is about noticing and understanding others' emotions and surroundings.

2-Correct answer: B
Explanation: Empathy is the ability to feel or understand what someone else might be experiencing.

3-Correct answer: B
Explanation: Emotional empathy is when you feel what another person is feeling.

4-Correct answer: C
Explanation: Respect means giving others space to share and be heard—even if you disagree.

5-Correct answer: A
Explanation: Micro-connections are quick, meaningful moments—like a smile or kind word—that build trust.

Scoring Guide

Count how many you got right:

5 correct: Social Superpower
You're aware, kind, and tuned in to others.

3–4 correct: Growing in Awareness
You understand a lot—just keep practicing empathy and curiosity.

0–2 correct: Starting the Journey
You're learning. Stay open and observe the world around you—you'll grow fast.

Self-Reflection Questions:

Use these questions to look inward and grow outward.

1. When someone I care about is upset, how do I usually respond?

2. When was the last time I really listened to someone without trying to fix them?

3. Have I ever judged someone before hearing their story? What might I do differently next time?

4. What's something I used to believe about others that changed after I got to know them better?

5. Do I notice when someone around me feels left out or uncomfortable? What signs do I usually miss?

6. What's one small thing I could do this week to show kindness or respect to someone?

7. What's a moment when someone truly understood me? How did it feel?

8. Who is someone different from me that I'd like to understand better?

NOTES

CHAPTER 4

RELATIONSHIP SKILLS

Build Real Friendships,
Communicate with Care

Section 1:

What Are Relationship Skills

Relationship skills are the set of abilities that help you form strong, respectful, and healthy connections with others. They're not just about being popular or getting along—they're about creating real, meaningful bonds.

These skills help you:

• Make new friends and keep them

• Work well in teams

• Handle arguments without hurting others

• Communicate your thoughts clearly

• Support others and ask for support when needed

You may think some people are just naturally good at relationships—but the truth is, these are skills you can practice and improve like any other.

Section 2:

Making and Keeping Friends

Friends are like emotional oxygen. They help us feel accepted, valued, and less alone.

But real friendship isn't instant—it takes time, trust, and care.

How to Make Friends:

• **Start small**: A smile, a compliment, or a simple "Hi" can open the door

• **Be curious**: Ask questions. People love to talk about themselves. Find shared interests

• **Be welcoming**: Invite others to join you in class, lunch, or activities

• **Be kind**: Kindness is magnetic. People are drawn to those who make them feel good

How to Keep Friends:

• **Show up**: Be reliable. If you say you'll call or hang out, follow through

• **Be honest**: Say how you feel, but with care. Avoid gossip or hidden anger

• **Celebrate and support**: Cheer for your friends' wins, and be there when they're down

• **Apologize**: Everyone messes up. What matters is how you own it and make it right

Friendship is not about being perfect. It's about being consistent, present, and kind.

Section 3:

Handling Conflict the Healthy Way

Even the best relationships hit rough patches.

The key is not avoiding conflict—but learning how to deal with it without damage.

Unhealthy Reactions:

• **Yelling**: Only raises walls

• **Blaming**: Puts others on defense

• **Ghosting**: Avoids the issue and hurts both sides

• **Talking behind someone's back**: Breaks trust

Healthy Conflict Skills:

• **Pause before reacting**: Let your brain calm down. Respond, don't explode

• **Use "I" statements**: Instead of "You're mean," try "I felt hurt when you ignored me"

• **Listen to understand**: Don't just wait for your turn to talk—really try to hear the other side

• **Take space if needed**: It's okay to cool off. But always come back to resolve things

Conflict, when handled well, makes relationships deeper—not weaker.

Section 4:

Communication That Connects

Words are powerful. How you speak can either build connection—or create misunderstanding.

Great Communicators:

• Make eye contact (when comfortable)

• Use a calm tone and body language that matches the message

• Speak honestly but kindly

• Ask questions like "Can you help me understand?"

• Listen without interrupting

• Validate feelings: "That sounds tough. I hear you"

It's not about using fancy words—it's about being clear, respectful, and genuinely interested in understanding others.

Section 5:

Asking for Help and Being There for Others

Good relationships are two-way streets. That means:

• You give help and receive it

• You support others—but also let yourself be supported

When to Ask for Help:

• You're overwhelmed, anxious, or unsure what to do

• You feel alone with a problem

• You need advice, guidance, or just someone to listen

How to Ask:

• "Can I talk to you about something?"

• "I don't need fixing, just someone to listen."

• "I'm having a tough time—can you be there for me?"

How to Offer Help:

• Check in with friends regularly

• Say, "I noticed you've been quiet—want to talk?"

• Just sit beside someone when words feel hard

Support builds trust. And trust builds strong relationships.

Section 6:

You Deserve Healthy Relationships

A healthy relationship feels like:

• You can be yourself

• You feel safe and not judged

• You're treated with kindness and honesty

• You're allowed to speak, say no, and be heard

If someone constantly makes you feel:

• Unsafe

• Small or wrong

• Afraid to speak up

• Drained, anxious, or confused

That relationship might be unhealthy. And it's okay to step back.

You deserve to be around people who:

• See your worth

• Respect your boundaries

• Want the best for you—even when things get hard

You don't need lots of friends. You just need the right ones.

Healthy Relationship Garden

Communication & Trust

Unhealthy Relationship Garden

Conflict & Mistrust

Neglect & Resentment

Trivia

Oxytocin Builds Bonds

• Oxytocin, known as the "connection hormone," is released when you hug, laugh, or have a deep conversation—it makes people trust and feel close to each other.

• The vagus nerve plays a role in feeling calm and connected. Just hearing someone's supportive tone can activate it.

• According to Harvard research, strong relationships are the #1 predictor of happiness over a lifetime—even more than wealth or fame.

Coming Next: Responsible Decision-Making

In Chapter 5, you'll learn:

• How to make choices that match your values

• How to think through consequences before you act

• How to trust your inner compass—not just peer pressure or impulses

Because healthy relationships grow stronger with healthy decisions.

Chapter 4 Quiz:

How Strong Are Your Relationship Skills?

Instructions:
Choose the best answer (A, B, or C).

1. What is the most important part of keeping a friendship strong?
A) Being interesting all the time
B) Being consistent, caring, and honest
C) Agreeing with your friend on everything

2. What's a healthy way to handle conflict?
A) Yell louder than the other person
B) Say nothing and never talk again
C) Use "I" statements and listen to understand

3. Which of these is a sign of poor communication?
A) Making eye contact and using kind words
B) Interrupting or ignoring someone when they talk
C) Asking questions to understand the other person

4. When should you ask for help in a relationship?
A) Only when you're about to give up
B) Anytime you feel overwhelmed or stuck
C) Never—you should always handle things alone

5. Which is a sign of a healthy relationship?

A) You're afraid to say how you really feel

B) You feel accepted, safe, and heard

C) You always say yes, even when you want to say no

Answers

1-Correct answer: B
Explanation: Strong friendships are built on trust, effort, and kindness—not on always being the same.

2-Correct answer: C
Explanation: Using "I felt..." helps express emotions without blame. Listening is key to peace.

3-Correct answer: B
Explanation: Interrupting or ignoring breaks connection. Great communicators listen.

4-Correct answer: B
Explanation: Asking for help early is a strength, not a weakness. It builds trust.

5-Correct answer: B
Explanation: Healthy relationships make you feel seen and valued—not afraid or trapped.

Scoring Guide

Count how many you got right:

5 correct: Connection Champion
You're growing strong, healthy relationships with wisdom and care.

3–4 correct: Relationship Builder
You're learning and doing well—keep practicing good communication and boundaries.

0–2 correct: Time to Tune In
It's okay. Now you know what to work on—and you've already started by learning.

Self-Reflection Questions:

These questions help you think more deeply about your own relationships.

1. What do I look for in a good friend—and do I try to be that kind of friend too?

2. How do I usually react when someone disagrees with me or hurts my feelings?

3. When was the last time I had a conflict? Did I react or respond? What would I do differently now?

4. Who in my life makes me feel safe, respected, and understood?

5. Is there a relationship in my life that drains me or makes me feel small? Why do I keep it?

6. Do I feel comfortable asking for help? If not, what stops me?

7. What's one communication habit I'd like to improve (e.g., listening more, being honest, using a calm voice)?

8. What's one thing I can do this week to strengthen a friendship or build a new one?

NOTES

CHAPTER 5

RESPONSIBLLE DECISION-MAKING

Choose Wisely, Live Bravely

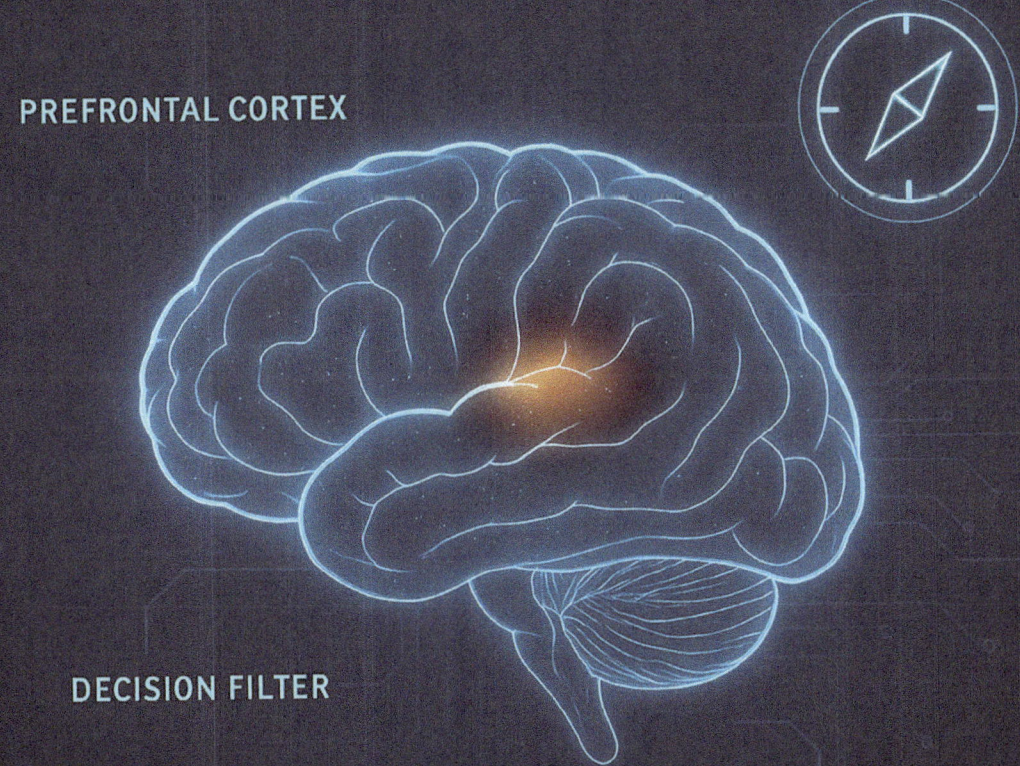

PREFRONTAL CORTEX

DECISION FILTER

DECISION FILTER

PROTOCOL ACTIVATED

Section 1:

What Is Responsible Decision-Making

Every day, you make choices.

Some are small: What do I wear? Should I scroll or sleep?

Some are big: Should I stand up for someone? Should I say yes or no?

Responsible decision-making means thinking things through, choosing based on values—not just feelings—and owning the results.

It's not about being perfect. It's about being thoughtful and brave enough to pause.

Section 2:

How Your Brain Makes Choices

Your brain has two systems:

• **Fast brain (amygdala):** Reacts quickly, based on fear, pleasure, or habits

• **Slow brain (prefrontal cortex):** Thinks ahead, solves problems, considers outcomes

When you act without thinking, your fast brain is in charge.

When you pause and reflect, your slow brain steps in.

Good decisions happen when you give your slow brain time to lead.

Section 3:

The 4-Step Decision-Making Tool

Here's a simple tool you can use any time you face a decision:

1. Pause – What's really going on? What's the choice here?

2. Think Ahead – What might happen next? (Best and worst case)

3. Check Your Values – What matters most to me right now?

4. Choose and Own It – Make your choice, and accept the result

You don't need to be 100% sure. You just need to be 100% present.

Section 4:

Feelings vs. Values

Feelings are real—but they don't always lead to the best actions.

You might feel:

- Like yelling

- Like giving up

- Like going along just to fit in

But ask yourself:

- Does this action match the kind of person I want to be?

- Will I feel proud of this choice later?

- Is this pressure—or is it truly me?

Your values are your inner compass. Let them lead.

Section 5:

Peer Pressure, Stress and Quick Decisions

Sometimes it feels like you have no time to think:

• Everyone's watching

• You feel embarrassed, scared, or angry

• You're tired or overloaded

That's when bad decisions happen.

What helps:

• Taking a breath

• Saying, "Let me think about it."

• Walking away—even just for a minute

Real strength is being able to pause when others are rushing.

Section 6:

Every Choice Shapes Your Life

Your decisions are like drops of water. One drop may feel small, but together they shape the river of your life.

* Saying no to something unsafe = **courage**

* Choosing kindness = **strength**

* Owning a mistake = **maturity**

* Standing alone for what's right = **leadership**

You won't always make perfect choices. No one does.

But with practice, you can build the habit of pausing, thinking, and choosing better every time.

Choose-Your-Path Forest

EASY WAY

COURAGE WAY

Trivia

The Brain Has a Decision Filter

• The prefrontal cortex (PFC) is your decision HQ—it's where logic, risk evaluation, and values come together.

• Teen brains are still developing their PFC, which is why teens are more likely to take risks without pausing.

• Studies show that when people "name their values" before making a decision, they're less likely to give in to peer pressure.

Coming Next: Problem Solving

In Chapter 6, you'll learn:

• How to break down problems step-by-step

• How to stay calm and focused when things go wrong

• How to turn mistakes into learning moments

Because decision-making starts the journey—but problem-solving helps you keep going when life gets tricky.

Chapter 5 Quiz:

How Thoughtful Are Your Decisions?

Instructions:
Choose the best answer (A, B, or C).

1. What does "responsible decision-making" mean?
   A) Making the fastest choice possible
   B) Doing what feels good right now
   C) Thinking ahead, using values, and owning the result

2. Which part of your brain helps with smart, calm decisions?
   A) Amygdala
   B) Brainstem
   C) Prefrontal Cortex

3. What's a good thing to do when you feel pressured or over-whelmed?
   A) Do what everyone else is doing
   B) React quickly before thinking
   C) Pause and say, "Let me think about it"

4. What is the purpose of checking your values before making a decision?
A) To make sure you're not breaking a rule
B) To win an argument
C) To make sure your choice reflects the kind of person you want to be

5. Which of these is a responsible decision?
A) Saying no to something unsafe, even if friends disagree
B) Doing what you're told, even if it hurts someone
C) Avoiding all decisions and letting others choose for you

Answers

1-Correct answer: C
Explanation: Responsible decision-making means thinking it through and choosing what aligns with your values—not just your emotions.

2- Correct answer: C
Explanation: The prefrontal cortex helps you plan, think through outcomes, and make thoughtful choices.

3-Correct answer: C
Explanation: Pausing helps your thinking brain (PFC) take charge instead of letting pressure or panic control you.

4-Correct answer: C
Explanation: Values are your inner compass—they help guide you even when emotions are strong.

5-Correct answer: A
Explanation: Choosing safety and self-respect—even when it's hard—is the heart of responsible decision-making.

Scoring Guide

Count how many you got right:

5 correct: Mindful Master
You're learning to think, pause, and lead yourself wisely.

3–4 correct: Thoughtful Thinker
You've got strong awareness—keep sharpening those tools.

0–2 correct: Learning to Pause
It's okay to start here. Reflection is the first step toward better choices.

Self-Reflection Questions:

Use these questions to explore how you make decisions—and how you can improve them.

1. What's a recent decision I made too fast? How might I have slowed down?

2. What's a choice I made recently that I'm proud of? Why?

3. When I feel pressured, do I usually go with the group, freeze, or speak up?

4. What are three values that matter most to me? (Examples: honesty, kindness, courage, fairness, freedom, family)

5. How do I usually feel after following my values—vs. ignoring them?

6. Who helps me pause and think clearly when I'm confused?

7. What's one area in life where I want to start making better decisions? (Friends, school, time, social media, emotions?)

8. What's a personal "pause phrase" I could use when I feel rushed? (Examples: "Give me a minute," "I need time to think," "Let me check with myself.")

NOTES

CHAPTER 6

PROBLEM SOLVING

Break It Down, Find a Way

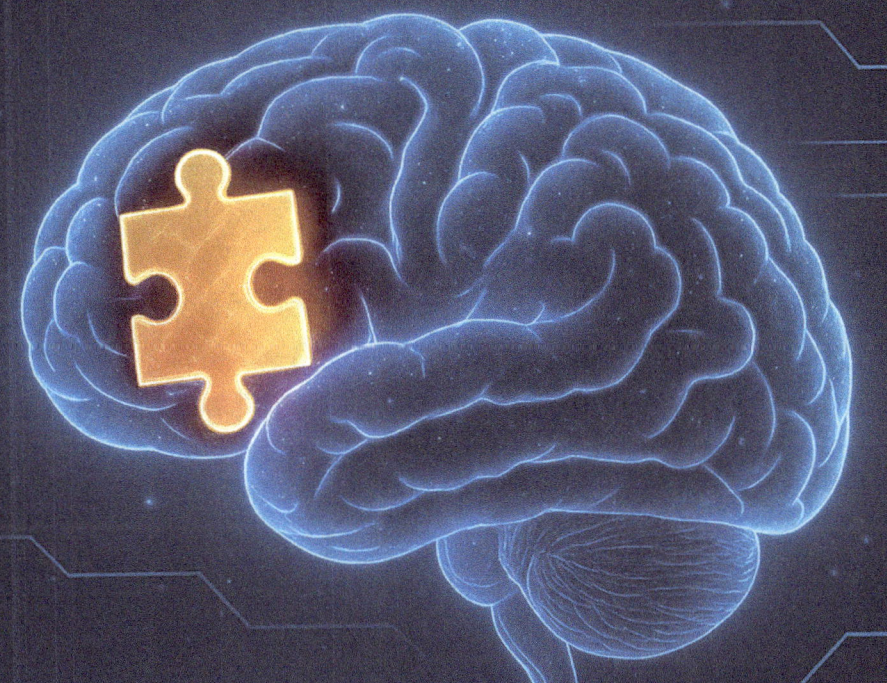

PROBLEM-SOLVING
MODE

Break It Down,
Find a Way

Section 1:

What Is Problem Solving

Problems happen. Every day.

Some are big: "My best friend isn't talking to me."

Some are small: "I forgot my homework."

Problem solving means facing the challenge instead of avoiding it—and finding a way forward, even if it's not perfect.

You don't need to fix everything instantly. You just need to take the next right step.

Section 2:

Why Your Brain Freezes (or Freaks Out)

When a problem shows up, your brain may:

• Freeze (feel stuck, numb, or overwhelmed)

• Flight (run away, avoid, procrastinate)

• Fight (get angry, defensive, reactive)

This is your survival brain taking over.

But problem-solving uses your thinking brain—the prefrontal cortex.

So first, calm down.
Then, think clearly.

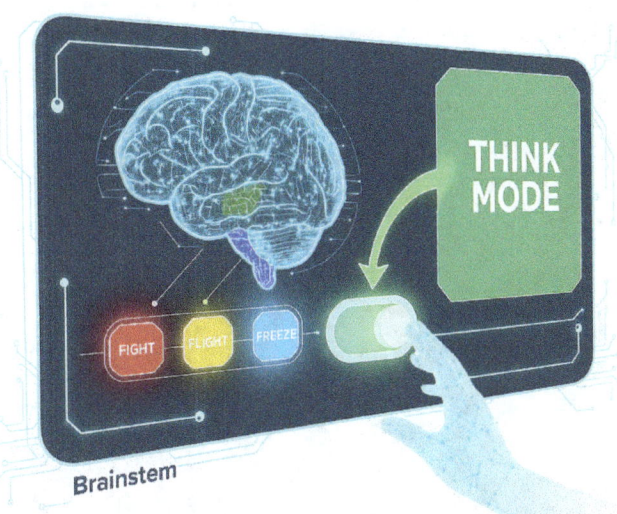

Brainstem

Section 3:

The 5-Step Problem-Solving Plan

Use these steps when facing any challenge:

1. Define the Problem:

What's really going on? Try to describe it in one sentence.

2. Explore Options:

What could I do? Brainstorm 3 possible actions—even if they're small.

3. Weigh the Pros and Cons:

What are the good and not-so-good things about each choice?

4. Choose One:

Pick the best option for now. You can always adjust later.

5. Try It and Reflect:

Take action. Then ask: Did it help? What would I do differently next time?

It's okay to try. It's okay to fail. What matters is learning and moving forward.

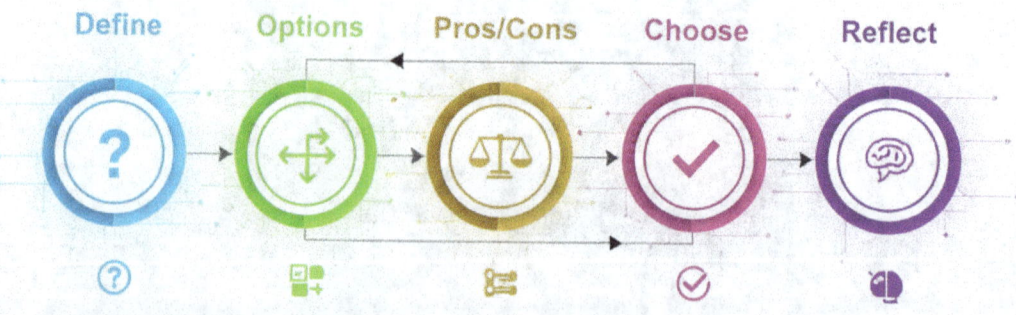

Section 4:

Examples of Everyday Problem Solving

Problem: You forgot your project deadline.

- **Define**: I missed a school deadline.

- **Options**: Ask for an extension, do it late anyway, give up.

- **Pros/Cons**: Extension = honest, effort; Late = less points; Giving up = regret.

- **Choose**: Ask for extension.

- **Reflect**: I should check deadlines weekly now.

Problem: Your friend is ignoring your texts.

- **Define**: My friend hasn't replied for 3 days.

- **Options**: Send one honest message, confront them, ignore them back.

- **Pros/Cons**: Honest message = clarity; Confront = drama; Ignore = more confusion.

- **Choose**: Send a kind message.

- **Reflect**: I can control how I reach out, not how they respond.

Section 5:

Mistakes = Growth

Mistakes don't mean failure. They mean learning.

Instead of asking, "Why did this happen to me?"

Ask: What can I learn from this?

Each time you solve a problem—even a tiny one—you build:

- **Confidence**

- **Flexibility**

- **Resilience**

Every problem solved makes your brain a little stronger.

Trivia

Problems Trigger Survival Circuits

• When you face a challenge, your brain's first response is survival—amygdala activation can make you freeze, flee, or fight.

• Naming the problem activates the prefrontal cortex and helps shift your brain from panic to planning.

• Problem-solving skills are a key part of executive functioning—the same brain set used for planning, organization, and time management.

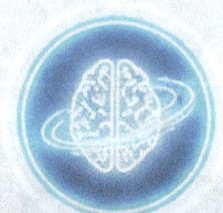

Coming Next: Cognitive Behavioral Techniques (CBT)

In Chapter 7, we'll explore:

• How your thoughts affect your feelings and actions

• How to challenge negative thinking

• How to build healthier mental habits

Because solving problems on the outside starts with thinking clearly on the inside.

Chapter 6 Quiz:

Are You a Problem Solver?

Instructions:
Choose the best answer (A, B, or C).

1. What is the first thing to do when you face a problem?
A) Panic and react quickly
B) Avoid it until it goes away
C) Define the problem clearly

2. What part of the brain helps with smart, calm problem-solving?
A) Brainstem
B) Amygdala
C) Prefrontal Cortex

3. What is the benefit of listing multiple solutions to a problem?
A) You confuse yourself more
B) You can pick the one that sounds easiest
C) You increase your chances of finding something that works

4. What should you do after trying a solution?
A) Forget about it and move on
B) Judge yourself if it didn't work
C) Reflect on what worked and what you'd do differently

5. What does solving problems regularly help build in your brain?

A) Anger and frustration

B) Resilience and confidence

C) More stress

Answers

1-Correct answer: C

Explanation: Naming the real problem helps your brain focus on solutions instead of staying overwhelmed.

2-Correct answer: C

Explanation: The prefrontal cortex helps you think logically, weigh options, and make a plan.

3-Correct answer: C

Explanation: Brainstorming gives you flexibility and control—it's okay if your first idea isn't perfect.

4-Correct answer: C

Explanation: Reflecting turns experience into wisdom. Every attempt teaches you something.

5-Correct answer: B

Explanation: Facing problems (even small ones) strengthens your brain's flexibility and emotional strength.

Scoring Guide

Count how many you got right:

5 correct: Master Problem Solver
You're learning how to stay calm, think clearly, and take action.

3–4 correct: Solution Starter
You've got the right tools—just keep practicing the steps.

0–2 correct: Growth in Progress
You're learning. Every small win builds your confidence. Keep trying.

Self-Reflection Questions:

Use these to apply what you've learned to your real life.

1. What's one small or big problem I'm facing right now?

2. When I feel overwhelmed, what do I usually do—freeze, fight, or avoid?

3. Can I describe the problem clearly in one sentence?

4. What are 2–3 possible solutions I haven't tried yet?

5. What's the worst thing that might happen if I try one? What's the best thing?

6. What helps me calm down before I can think clearly?

7. When was the last time I solved a tough problem—and how did it feel?

8. What's one daily habit I can build to face problems instead of ignoring them?

NOTES

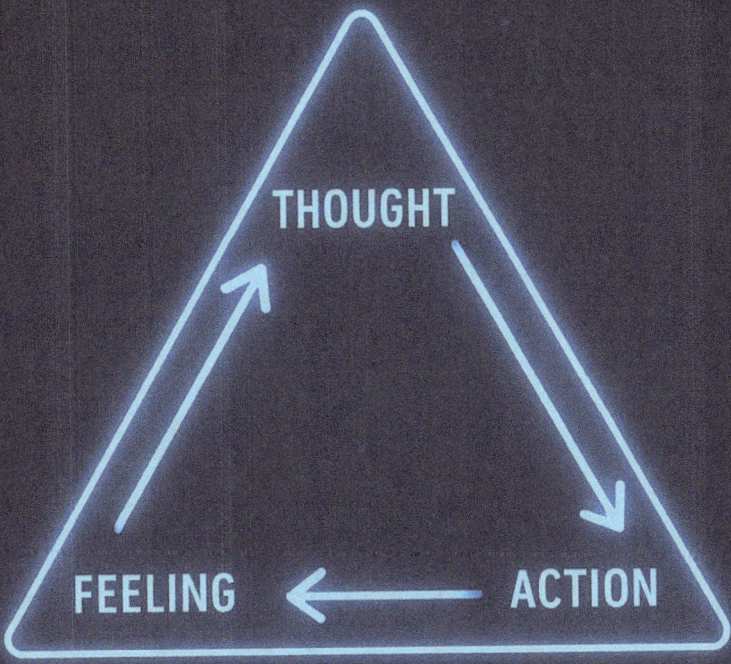

COGNITIVE BEHAVIORAL TECHNIQUES (CBT)

TRAIN YOUR THOUGHTS, CHANGE YOUR MIND

FOR AGES 13+ AND CURIOUS ADULTS

Section 1:

What Is CBT

Cognitive Behavioral Techniques (CBT) are practical skills that help you:

• Notice your thoughts

• Understand how they affect your feelings and actions

• Change thoughts that are unhelpful or untrue

It's one of the most researched and effective tools used by psychologists around the world.

The core idea is simple:

Thoughts -> Feelings -> Actions

Change your thoughts, and you change your life.

Section 2:

The Thought-Feeling-Action Triangle

Imagine a triangle:

• One corner is thoughts (what you tell yourself)

• One corner is feelings (what you feel in your body or emotions)

• One corner is actions (what you do)

They're all connected.

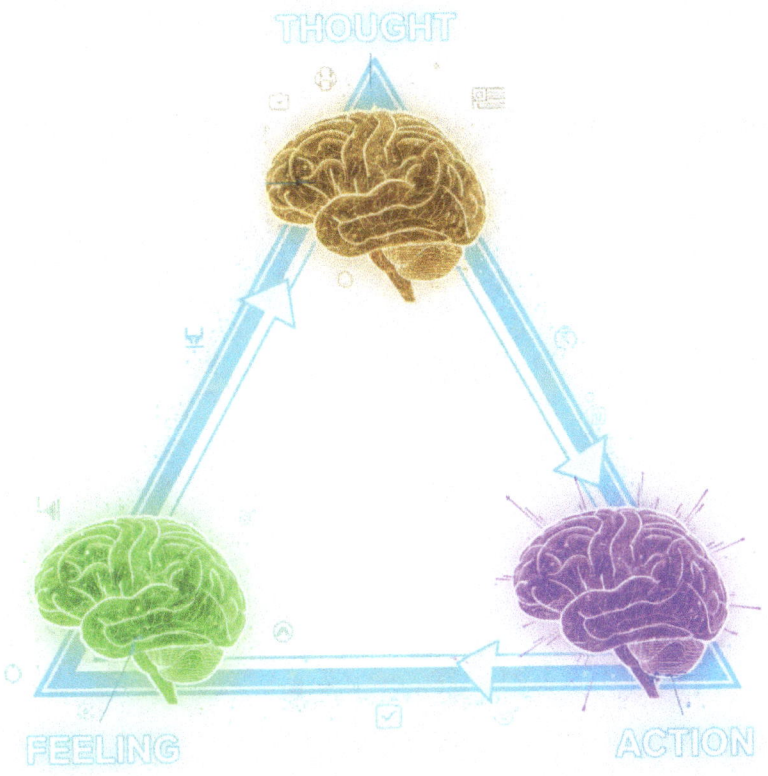

Example:

- Thought: "Nobody likes me."

- Feeling: Sad, anxious

- Action: Withdraw, avoid people

Now flip it:

- New Thought: "Some people care about me."

- New Feeling: Hopeful

- New Action: Text a friend, smile back, show up

You can't always control feelings. But you can train your thoughts.

Section 3:

Spotting Thinking Traps

Your brain plays tricks on you sometimes. Here are common thinking traps to watch for:

1. All-or-Nothing Thinking: "I failed once, so I'll always fail."

2. Mind Reading: "She didn't text back—she must be mad at me."

3. Catastrophizing: "If I mess up, everything will fall apart."

4. Labeling: "I'm stupid. I'm lazy."

5. Shoulds and Musts: "I should never make mistakes."

These thoughts feel real—but they aren't always true.

Step 1: Catch the thought.

Step 2: Challenge it.

Section 4:

How to Challenge a Thought

Here's how you flip a negative thought:

1. **Ask**: Is this 100% true? Or just how I feel right now?

2. **Ask**: What would I tell a friend who thought this?

3. **Ask**: What's a more helpful or balanced way to think?

Example:

Old Thought: "I'll never be good at this."

New Thought: "I'm still learning. Progress takes time."

This isn't fake positivity. It's training your brain to be fair, not fearful.

Thought Filtering

Negative Thought enters

Balanced Thought exits

Section 5:

Practice Makes Pathways

Your brain works like a muscle. The more you practice a thought, the stronger that pathway becomes.

Every time you:

• Pause and notice your thoughts

• Replace harsh thoughts with realistic ones

• Take one action based on hope instead of fear

You rewire your brain for resilience.

CBT is like brain fitness. The more you do it, the easier it becomes.

Section 6:

CBT and Mental Health

CBT is used to help with:

- Anxiety

- Depression

- Anger

- Self-doubt

- Stress

It doesn't mean your feelings are wrong. It means you're learning how to guide your mind with skill—not shame.

CBT isn't about ignoring your emotions. It's about becoming the driver of your mind—not the passenger.

Trivia

Your Brain Believes Your Thoughts

• Thoughts can change your brain's chemistry! Positive, realistic thoughts increase serotonin and dopamine—your mood boosters.

• CBT works by identifying "thinking traps" like catastrophizing, all-or-nothing thinking, and mind-reading—then challenging them to rewire your response.

• Neuroplasticity studies show that repeated CBT practices physically rewire your brain's connections—creating new default pathways for optimism and action.

Coming Next: Positive Psychology

In Chapter 8, you'll learn:

• How to focus on what's going right in your life

• How to build gratitude, strengths, and joy

• How to create a mindset of growth and positivity

Because it's not just about managing the tough stuff—it's about growing the good.

Chapter 7 Quiz:

Are You the Boss of Your Thoughts?

Instructions:
Choose the best answer (A, B, or C).

1. What does CBT help you do?
A) Control other people's behavior
B) Ignore your feelings completely
C) Notice, understand, and change unhelpful thoughts

2. What are the three parts of the thought-feeling-action triangle?
A) Breathing, sleeping, exercising
B) Thoughts, feelings, actions
C) School, friends, family

3. Which of these is a thinking trap?
A) "It's okay to make mistakes."
B) "I'll try again tomorrow."
C) "I failed once, so I'll always fail."

4. What is the first step in challenging a negative thought?
A) Believe it without question
B) Ask yourself, "Is this 100% true?"
C) Try to forget it immediately

5. What happens when you practice realistic, helpful thoughts regularly?
A) Your brain gets more stressed
B) Your brain creates stronger positive pathways
C) Nothing changes

Answers

1-Correct answer: C
Explanation: CBT teaches you to identify unhelpful thinking patterns and change them to improve your emotions and actions.

2-Correct answer: B
Explanation: The triangle shows how your thoughts affect how you feel and what you do.

3-Correct answer: C
Explanation: That's an example of all-or-nothing thinking—a common mental trap.

4-Correct answer: B
Explanation: You start by questioning the accuracy of your thought with curiosity, not judgment.

5-Correct answer: B
Explanation: Like muscles, the more you practice helpful thinking, the stronger those brain circuits become.

Scoring Guide

Count how many you got right:

5 correct: Thought Ninja
You're learning to take control of your mind like a true CBT champion.

3–4 correct: Mental Rewiring in Progress
You're building great awareness—keep practicing and you'll get stronger every day.

0–2 correct: Start Noticing, Start Growing
Be proud you're learning this now. Everyone starts somewhere, and you've just taken your first step.

Self-Reflection Questions:

Use these questions to apply CBT to your everyday thoughts and emotions.

1. What's a negative thought I tell myself often?

2. How does that thought make me feel and act?

3. Is that thought always true—or just how I feel in the moment?

4. What would I say to a friend who had that thought?

5. What's a more realistic and helpful way I could think instead?

6. What's one thinking trap I fall into most often? (Examples: catastrophizing, mind reading, all-or-nothing thinking)

7. When was a time I changed my thought—and it helped me feel better?

8. What's one sentence I want to practice telling myself this week?

NOTES

CHAPTER 8

POSITIVE PSYCHOLOGY

Grow the Good, Strengthen Your Joy

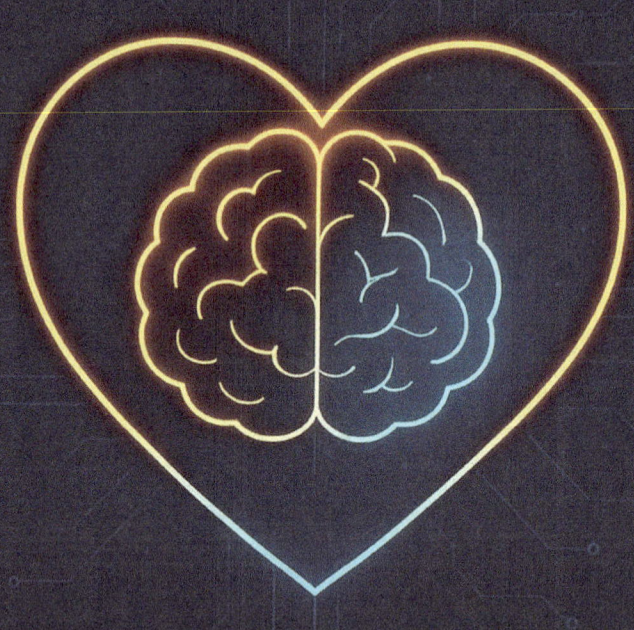

Section 1:

What Is Positive Psychology

Positive psychology is the science of what makes life worth living.

It's not about pretending everything is perfect. It's about growing what's strong—even when things feel hard.

Traditional psychology asks: "What's wrong?"

Positive psychology asks: "What's right, and how do we build more of it?"

It focuses on strengths, gratitude, purpose, joy, kindness, and hope.

FIELD OF POSITIVE EMOTIONS

Section 2:

The Power of Strengths

Everyone has natural strengths—ways of thinking, feeling, or acting that energize them.

Examples include curiosity, humor, kindness, leadership, creativity, bravery, and gratitude.

When you use your strengths daily, you feel more:

• Confident

• Engaged

• Purposeful

Knowing your strengths helps you thrive—even during hard times.

Section 3:

Gratitude Changes Your Brain

Gratitude isn't just saying "thank you." It's a mindset of noticing what's going right.

Practicing gratitude daily has been shown to:

• Increase happiness

• Lower stress

• Improve sleep

• Strengthen relationships

Try This:

• Write 3 good things that happened today—even small ones

• Say thank you to someone who made a difference

• Reflect: "What did I appreciate about today?"

Gratitude trains your brain to look for light—even in darkness.

Section 4:

Positive Emotions Build Resilience

When you experience emotions like joy, love, awe, or inspiration, your brain opens up.

Positive emotions help you:

- Think more clearly

- Solve problems better

- Recover faster from stress

This is called the broaden-and-build effect—your mind grows stronger the more you practice joy.

Find Micro-Moments:

- Laugh at something silly

- Smile at a friend

- Enjoy a sunset or song

Small joys fuel big strength.

Section 5:

Kindness and Purpose

Doing good feels good.

When you help others, your brain releases "helper's high" chemicals like dopamine and oxytocin.

Acts of kindness also:

• Boost your mood

• Lower anxiety

• Build social connection

Purpose means knowing that your life matters—and using your time, energy, and voice to make a difference.

Even small purposes, like cheering up one person, can give your life meaning.

Section 6:

Build a Positive Psychology Habit

To grow the good, practice these habits:

- **Daily Gratitude**: Write or say 3 things you're thankful for

- **Strength Check-In**: Ask, "What strength did I use today?"

- **Random Kindness**: Do something nice—no reward needed

- **Joy Journal**: Track moments that made you smile

- **Purpose Reminder**: Ask, "How can I help today?"

You don't have to wait to feel better. You can practice it daily—starting now.

Trivia

Gratitude Changes Your Brain

• Practicing gratitude activates the ventral striatum and medial prefrontal cortex—areas associated with joy and emotional regulation.

• People who list 3 things they're grateful for each day sleep better and report higher well-being, according to a University of California study.

• Acts of kindness increase oxytocin (bonding), dopamine (reward), and endorphins (feel-good chemicals)—creating a "helper's high."

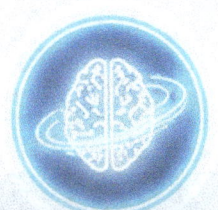

Coming Next: Mindfulness

In Chapter 9, you'll learn:

• How to stay calm in the moment?

• How to train your attention?

• How to reduce overthinking and increase presence?

Because joy grows deeper when your mind is here—not lost in the past or racing into the future.

Chapter 8 Quiz:

Are You Growing the Good?

Instructions:
Choose the best answer (A, B, or C).

1. What is the main focus of positive psychology?
A) Ignoring problems and pretending to be happy
B) Fixing all weaknesses as fast as possible
C) Growing strengths, gratitude, joy, and meaning

2. What are character strengths?
A) Physical abilities like running and jumping
B) Skills only adults have
C) Natural inner traits like kindness, creativity, bravery

3. How does gratitude affect your brain and mood?
A) It makes you ignore real problems
B) It reduces stress and improves happiness
C) It only works if big things happen

4. What is the "broaden-and-build" effect?
A) When stress helps you study harder
B) When positive emotions help your brain grow stronger
C) When you get better at arguing

5. What happens when you show kindness to others?
A) You feel more tired and drained
B) People take advantage of you
C) Your brain releases feel-good chemicals and you feel more connected

Answers

1-Correct answer: C
Explanation: Positive psychology builds what's strong, not just repairs what's wrong.

2-Correct answer: C
Explanation: Everyone has character strengths—they just need to recognize and use them.

3-Correct answer: B
Explanation: Gratitude rewires your brain to notice what's working—even in tough times.

4-Correct answer: B
Explanation: Joy, love, and awe open your mind, helping you think clearly and build resilience.

5-Correct answer: C
Explanation: Helping others boosts dopamine and oxytocin—building happiness and connection.

Scoring Guide

Count how many you got right:

5 correct: Joy Builder
You're living with intention and training your mind to grow what's good.

3–4 correct: Positive In Progress
Great work. Keep using strengths, gratitude, and kindness every day.

0–2 correct: Starting to Shine
You're learning the tools to build joy—and that's the first step to true strength.

Self-Reflection Questions:

These help you explore your strengths and grow the joy inside you.

1. What's one strength I have that makes me feel energized when I use it?

2. What's something small I appreciated about today?

3. When I feel stressed, what positive emotion could I try to create instead? (Examples: curiosity, humor, kindness)

4. What's one act of kindness I did recently—and how did it make me feel?

5. Who is someone I'm grateful for, and have I told them?

6. What's something I enjoy that makes me lose track of time (a "flow" activity)?

7. What gives me a sense of purpose—even in small ways?

8. How can I build one daily habit to grow my joy? (Examples: gratitude journal, compliment a friend, take a nature break)

NOTES

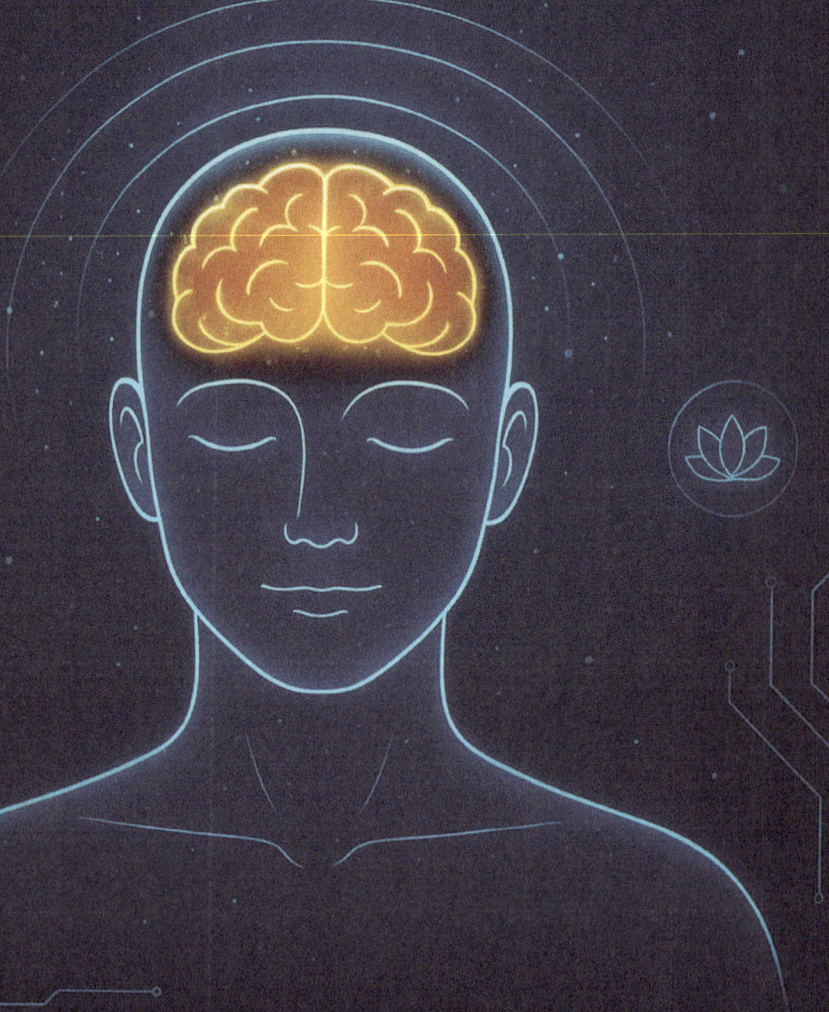

CHAPTER 9
MINDFULNESS
Be Here Now

Section 1:

What Is Mindfulness

Mindfulness means paying full attention to the present moment—on purpose, without judgment.

It's the opposite of:

• Living on autopilot

• Overthinking the past

• Worrying about the future

When you're mindful, you're here—not lost in your head. You're aware of your thoughts, body, breath, and surroundings.

Mindfulness doesn't mean emptying your mind. It means watching it with kindness.

Section 2:

Why Mindfulness Helps

Practicing mindfulness can:

- Reduce stress and anxiety

- Improve focus and memory

- Help with sleep and mood

- Calm your body and nervous system

Your brain has a Default Mode Network (DMN) that wanders and overthinks. Mindfulness helps quiet that network and builds your prefrontal cortex—the part that keeps you present.

Section 3:

The Power of the Breath

Your breath is always with you.
It connects your body and mind.
It's a free, portable calming tool.

Try This:
1-Minute Breathing Practice

- Sit or stand comfortably
- Breathe in slowly for 4 seconds
- Breathe out slowly for 6 seconds
- Repeat for one minute, watching your breath go in and out

Notice your thoughts, but return to your breath. Every time you come back—you're strengthening your focus.

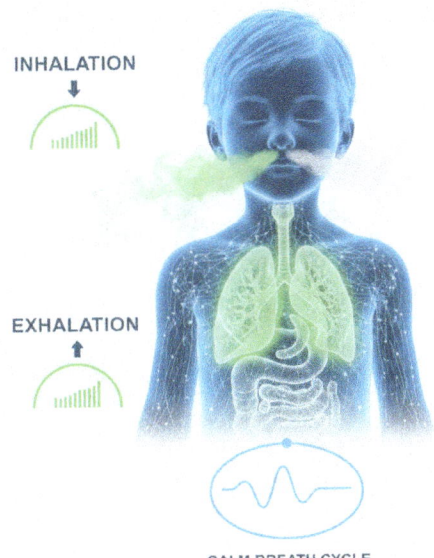

INHALATION

EXHALATION

CALM BREATH CYCLE

Section 4:

Common Myths About Mindfulness

"I can't do it. My mind is too busy."
Busy minds are normal. The point is to notice, not control.

"It's only for calm people."
No. It's for stressed people, anxious people, distracted people—everyone.

"It takes hours a day."
Even 30 seconds of mindful breathing makes a difference.

Mindfulness is not perfect focus—it's gentle returning.

Section 5:

Everyday Mindfulness

You don't have to sit on a mountain to practice.

You can be mindful during:

* Eating (notice taste, texture, smell)

* Walking (feel your feet, hear sounds around you)

* Showering (sense the water, your breath, your body)

* Talking (truly listen instead of planning your reply)

Any moment can be a mindfulness moment.

Section 6:

The Mindful Mindset

Mindfulness isn't just a practice—it's a way of living.

It means approaching life with:

- **Curiosity:** What's really happening here?

- **Compassion:** Especially toward yourself

- **Non-judgment:** This is what I feel. It's okay.

- **Presence:** I am here. I am safe. I am enough.

The more you practice, the easier it gets to come back to now.

Trivia

Mindfulness Shrinks the Amygdala

• Long-term mindfulness practice has been shown to reduce the size and reactivity of the amygdala (fear center).

• Mindfulness strengthens the prefrontal cortex and hippocampus—boosting memory, focus, and emotional regulation.

• Just 5 minutes of mindful breathing a day has been shown to reduce cortisol levels and improve mood in teenagers.

Coming Next: Resiliency

In Chapter 10, you'll learn:

• How to bounce back after setbacks

• How to train your brain for emotional recovery

• How to build inner strength that lasts

Because presence creates peace—but resilience turns it into power.

Chapter 9 Quiz:

How Mindful Are You?

Instructions:
Choose the best answer (A, B, or C).

1. What does mindfulness mean?
A) Emptying your mind completely
B) Focusing only on positive thoughts
C) Paying full attention to the present moment, without judgment

2. Which part of the brain becomes stronger with mindfulness?
A) Brainstem
B) Prefrontal Cortex
C) Amygdala

3. What is the best way to handle a "busy mind" during mindfulness?
A) Get frustrated and stop trying
B) Judge yourself for not being focused
C) Gently notice the thought and return to your breath

4. Which of the following is a form of mindfulness?
A) Listening closely while a friend is speaking
B) Multitasking while scrolling your phone
C) Daydreaming during a conversation

5. How long do you have to practice mindfulness for it to help?
A) At least 2 hours a day
B) Only during yoga or meditation retreats
C) Even 30 seconds of mindful breathing can make a difference

Answers

1-Correct answer: C
Explanation: Mindfulness is about being aware of what's happening right now—inside and around you—with curiosity and kindness.

2-Correct answer: B
Explanation: The prefrontal cortex (the thinking, focusing part) becomes stronger with consistent mindfulness practice.

3-Correct answer: C
Explanation: Mindfulness is not about perfect focus—it's about gently coming back whenever your mind wanders.

4-Correct answer: A
Explanation: Mindfulness includes being fully present while listening—without distraction or judgment.

5-Correct answer: C
Explanation: Short, consistent mindfulness practices help calm your brain and improve focus—even just 30 seconds at a time.

Scoring Guide

Count how many you got right:

5 correct: Mindfulness Master
You're present, focused, and building a calm mind daily.

3–4 correct: Aware and Growing
You understand mindfulness well—just keep practicing.

0–2 correct: Beginning the Journey
That's okay. Mindfulness is a skill, not a talent. You're already doing great by learning it.

Self-Reflection Questions:

Use these to explore how mindfulness shows up in your life—and how to grow it.

1. What usually pulls my mind away from the present moment? (Examples: overthinking, stress, social media)

2. What does "being present" feel like in my body and breath?

3. When was a moment today that I felt truly calm or focused?

4. How do I usually respond to stress—and what would it feel like to pause instead?

5. What's one daily activity I could do more mindfully? (Examples: brushing teeth, eating lunch, walking, listening)

6. When my thoughts feel overwhelming, what can I gently return to? (Example: my breath, a sound, my body)

7. What would happen if I stopped judging my thoughts—and just noticed them?

8. What's one sentence I want to tell myself when I feel distracted or anxious? (Examples: "Come back to now." "I am safe." "Just breathe.")

NOTES

CHAPTER 10
RESILIENCY
Bounce Back, Grow Stronger

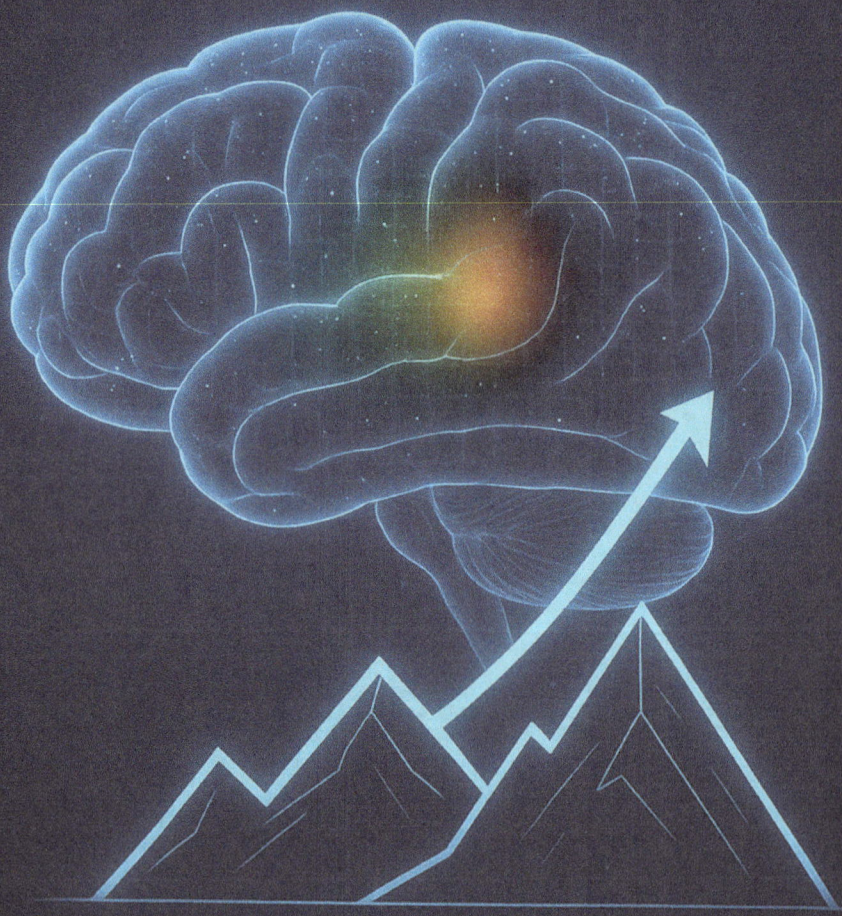

RESILIENCE PROTOCOL ACTIVATED

Section 1:

What Is Resiliency

Resiliency is your ability to bounce back from stress, failure, or hard times—and grow stronger because of them.

It doesn't mean being tough all the time.

It means learning to bend without breaking and believing, "I can handle this. I can grow from this."

Resilience is not what happens to you—it's how you respond.

Section 2:

Stress Is Not the Enemy

Stress is part of life. It's not always bad.

• Some stress helps you grow (like a workout for your brain)

• Too much stress can overload you

• No stress at all can leave you bored or unmotivated

The key is how you recover.

Resilient people know how to rest, reflect, and recharge.

Section 3:

The Resilient Brain

Resilience lives in your nervous system.

When you face a challenge, your brain and body respond with:

- Alertness (heart races, thoughts speed up)

- Emotion (fear, sadness, frustration)

- Urge to act (fight, flight, freeze)

Resilience means knowing how to:

- Calm your system

- Soothe your body

- Regain clarity before reacting

The stronger your prefrontal cortex, the better your bounce-back.

Section 4:

Bounce-Back Tools

Here are simple tools to build daily resilience:

1. Name It to Tame It
Say how you feel. This activates the thinking brain and calms the emotion brain.

2. Body Reset
Breathe, stretch, splash water, take a walk. Calm the body = calm the brain.

3. Reframe the Story
Ask: "What's another way to see this?" or "What is this teaching me?"

4. Ask for Support
You're not meant to bounce alone. Reach out.

5. Rest with Intention
Sleep, nature, silence, art. Choose what fills you—not just numbs you.

Resilient people don't avoid emotions. They face them with tools.

Section 5:

Growth After Struggles

Challenges can create change.

Many people report becoming:

• Wiser

• More compassionate

• Clearer about what matters

• More mentally strong

This is called post-traumatic growth.

Instead of asking, "Why me?" resilient people ask:

"What now?" or "How can I grow through this?"

Section 6:

Resiliency Is a Daily Practice

You build resilience like a muscle—one day at a time.

- When you get back up after a tough moment
- When you speak kindly to yourself instead of giving up
- When you pause instead of panic

You're not weak for struggling. You're growing through the stretch.
The strongest people aren't always loudest—they're the ones who keep showing up.

RESILIENCE MOUNTAIN TRAIL

SUMMIT: HIGHER PERSPECTIVE

STARTING POINT: COMFORT VALLEY

DIFFICUITY → ELEVATION

Trivia

Stress Can Build Strength—If You Recover

• Moderate stress followed by healthy recovery builds resilience—it's called the "Stress Inoculation Effect."

• Post-traumatic growth is a real, researched phenomenon: after adversity, people often report greater appreciation for life, better relationships, and stronger inner strength.

• Resilience is linked to high vagal tone—the ability of your nervous system to bounce back quickly after a challenge.

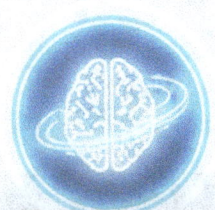

Chapter 10 Quiz:

How Resilient Are You?

Instructions:
Choose the best answer (A, B, or C).

1. What does resiliency mean?
A) Never feeling stress or pain
B) Hiding your emotions to look strong
C) Bouncing back from tough times and learning from them

2. What helps your brain recover after stress?
A) Ignoring the problem
B) Judging yourself for being upset
C) Calming your body and naming what you feel

3. What is Post-Traumatic Growth?
A) A diagnosis for anxiety
B) A fake idea used in therapy
C) Positive growth that comes after hard experiences

4. Which of the following is a bounce-back tool?
A) Reframing the story ("What is this teaching me?")
B) Pretending nothing happened
C) Reacting quickly and emotionally

5. What's a sign that someone is practicing resilience daily?
A) They never cry or struggle
B) They keep showing up and trying, even when it's hard
C) They avoid all stressful situations

Answers

1-Correct Answer: C
Explanation: Resiliency is about recovery and growth—not avoiding stress, but learning how to respond.

2-Correct Answer: C
Explanation: Naming emotions and resetting your body helps engage the thinking brain and calm the stress response.

3-Correct Answer: C
Explanation: Many people report becoming stronger, wiser, and more compassionate after overcoming adversity.

4-Correct Answer: A
Explanation: Reframing helps you gain perspective and shift your brain from panic to growth.

5-Correct Answer: B
Explanation: True resilience shows up quietly—in persistence, self-compassion, and the willingness to try again.

Scoring Guide

Count how many you got right:

5 out of 5 – Resilience Rock Star

You understand how to bounce back with strength and clarity.

3 to 4 – Strong and Growing

You've built great tools—keep practicing, especially during small challenges.

0 to 2 – Just Beginning

Resilience is a skill. You're already on the path by learning and trying.

Self-Reflection Questions:

Use these to explore your own bounce-back mindset.

1. When I face something stressful, what's my first reaction?

2. How do I usually treat myself when I'm struggling? Supportive or self-critical?

3. What's one tool from this chapter I can use the next time I feel overwhelmed?

4. Think of a challenge I've faced in the past. What did I learn from it?

5. What does "bouncing back" look like for me—not perfectly, but realistically?

6. Who are the people I can reach out to when I need support?

7. What helps me rest, recharge, or reset my emotions in a healthy way?

8. What's one phrase I want to remember when I'm going through something tough?
Examples: "This will pass." "I can handle hard things." "I grow in the stretch."

NOTES

Completion Note

This completes the Core 10 Competencies Series.

Your brain is now trained in:
- Self-Awareness
- Self-Management
- Social Awareness
- Relationship Skills
- Responsible Decision-Making
- Problem Solving
- CBT Tools
- Positive Psychology
- Mindfulness
- Resiliency

You're ready to practice, grow, and share these skills with the world.

WHAT IS NEXT?

NEXT STEPS FOR MIGHTY CHAMPIONS

You are now trained in the Ten Core Competencies of Mental Health Education. Your journey doesn't stop here—this is where it begins.

Mighty Champions get access to real opportunities that strengthen your confidence, leadership, community impact, and future college or career profile.

Scan the QR code or visit the link to access your next steps.

1. Register for Community Volunteer Hours

Earn verified community-service hours by sharing what you've learned.

Examples of volunteer activities:
- Read a chapter of this book to a younger child or a group
- Teach one competency in your school club
- Lead a mindfulness minute in class
- Start a Mental Health Champion Club

You will receive official documentation for your hours.

2. Register for Research Participation

Become part of real student research in mental health education.

Opportunities include:
* Research posters
* Short papers

- Student presentations
- Speaking opportunities at student events
  Your work may be highlighted on social media, newsletters, or youth conferences.

3. Register for the Final Exam and Certification
Take the official Mighty Champions Certification Exam. After passing, you will receive:

* Mighty Champions of Mental Health Education Certificate
* Official transcript of completed competencies
* Documentation for your CV, resume, or college application
  This certification shows you are ready to support mental-health education in your school, home, and community.

4. Register to Participate in Our Podcast
Share your voice. Join TheRaghavPodcast as a guest.
You may:

- Discuss a chapter
- Share your personal growth
- Interview mental-health experts
- Present a research topic
  Your story can inspire thousands.

What you do next matters.

Use your skills. Teach others. Build community. Grow into the leader you were meant to be.

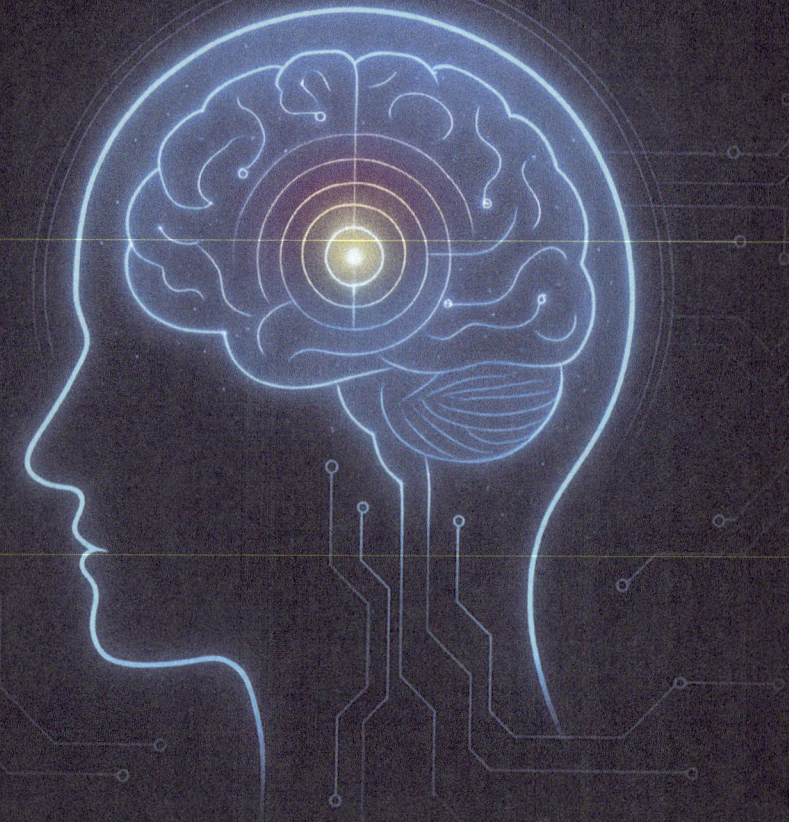

CHAPTER 11

METACOGNITION

Thinking About Thinking

Section 1:

What Is Metacognition?

Metacognition means "thinking about thinking." It's your brain's ability to observe itself.

When you pause and ask:

- Why did I just think that?

- How am I solving this problem?

- Is this thought helping me or hurting me?

You are using metacognition.

Scientists call it one of the most powerful tools for learning, mental health, and personal growth.

Metacognition helps you understand how your mind works—so you can guide it better.

Section 2:

The Brain Behind the Watching Brain

Metacognition happens in the prefrontal cortex—the front part of your brain responsible for:

- **Planning**
- **Self-control**
- **Decision-making**
- **Reflection**

Studies using brain imaging (fMRI) show that people who practice metacognition regularly have more activity in this area. That means better emotional control, smarter decisions, and stronger learning.

SELF-OBSERVATION CYCLE

Section 3:

Scientific Examples

Example 1: Math Problem Solving
Students who stop and ask, "How am I approaching this problem?" often solve it better. They're not just doing math—they're watching how they do math.

Example 2: Memory Recall
When learners reflect on which study methods work for them—like flashcards or writing things down—they remember more. This is the basis of tools like active recall and spaced repetition.

Example 3: Emotional Awareness
In therapy, people are taught to observe their thoughts. For example: "Is this a fact or just a habit of thinking?" This kind of self-observation helps people change negative patterns.

Section 4:

Why It Matters

Without metacognition, we react without thinking.

With metacognition, we respond with awareness.

It helps you:

• Notice negative thought loops

• Make smarter choices

• Learn and study more effectively

• Understand yourself better in relationships

It turns you from being just a thinker into the guide of your own mind.

Section 5:

How to Strengthen Metacognition

Here are five science-backed ways to grow your metacognition muscle:

1. Think Aloud – Talk through your thinking process. It's used in classrooms and therapy to improve clarity.

2. Reflect Journaling – Write down what you were thinking today, why, and how it affected your emotions or actions.

3. Ask Metacognitive Questions:

- What helped me succeed here?

- Where did this belief come from?

- What might I be avoiding with this thought?

4. Teach Someone Else – Teaching makes you more aware of how you think and learn.

5. Pause and Label – Notice a thought and name it: "That's a worry," "That's a memory," "That's fear talking."

Small steps like these rewire how your brain talks to itself.

Section 6:

Final Thought

You are not your thoughts—you are the one who notices them.

Metacognition is not about controlling your brain like a robot.

It's about understanding it like a scientist.

It's about stepping back like a wise friend and asking:

- **What's going on in my mind right now?**

- **Is this thought useful?**

- **What would a wiser version of me do next?**

That is metacognition.

And it's one of the most powerful, most human things your brain can do.

Chapter 11 Quiz:

Are You Thinking About Thinking?

Instructions:
Choose the best answer (A, B, or C).

1. What is metacognition?
A) Memorizing facts more efficiently
B) Controlling other people's thoughts
C) Thinking about your own thinking processes

2. Which part of the brain is most involved in metacognition?
A) Brainstem
B) Prefrontal Cortex
C) Amygdala

3. Which of the following is an example of metacognitive thinking?
A) Solving a math problem without checking your method
B) Wondering why you keep thinking negatively in stressful situations
C) Watching TV to distract your mind

4. Why do students who reflect on their learning perform better?
A) Because they memorize everything faster
B) Because they know how to ask better questions and fix mistakes
C) Because they copy smarter people

5. What's a simple way to build metacognitive awareness?
A) Think as fast as possible
B) Judge your thoughts quickly
C) Pause and label your thoughts as they arise s

Answers

1-Correct Answer: C
Explanation: Metacognition means observing and understanding how you think, learn, and respond.

2-Correct Answer: B
Explanation: The prefrontal cortex helps with reflection, decision-making, and monitoring your thoughts.

3-Correct Answer: B
Explanation: Metacognitive thinking involves questioning and reflecting on how you think or react.

4-Correct Answer: B
Explanation: Metacognition helps students adjust strategies, understand mistakes, and learn deeply.

5-Correct Answer: C
Explanation: Noticing and naming thoughts helps you separate from automatic patterns and build awareness.

Scoring Guide

Count how many you got right:

5/5 – Meta-Master
You're aware of your thinking—and becoming a guide for your own brain.

3–4 – Mind Explorer
You're reflecting well. Keep pausing and questioning what drives your thinking.

0–2 – Observer in Training
That's okay—just learning these questions puts you on the path to deeper self-awareness.

Self-Reflection Questions:

These questions help develop your metacognitive muscles—no rush, no right answers, just curiosity.

1. What kinds of thoughts show up most often in my mind each day? (Examples: worries, planning, comparing, criticizing, solving)

2. When I feel stuck or upset, do I ever stop and ask: "What am I thinking right now?"

3. What are three thought patterns I've noticed that either help me—or hold me back?

4. When I make a mistake, do I reflect on how I made that choice, or just blame myself?

5. What's a recent situation where I changed how I thought—and it helped me handle it better?

6. What would it look like if I paused and "labeled" my thoughts today? (e.g., "That's a judgment," "That's a fear," "That's a plan")

7. Who in my life seems good at thinking clearly—and what habits do they have?

8. If I taught someone younger than me how to use their mind better, what's one tip I'd share?

NOTES